Solar Power

Solar Power

Jaime Hunter

Callisto Reference

www.callistoreference.com

Callisto Reference,
118-35 Queens Blvd., Suite 400,
Forest Hills, NY 11375, USA

Visit us on the World Wide Web at:
www.callistoreference.com

ISBN: 978-1-64116-594-5 (Hardback)

Cataloging-in-Publication Data

Solar power / Jaime Hunter.
p. cm.
Includes bibliographical references and index.
ISBN 978-1-64116-594-5
1. Solar energy. 2. Renewable energy sources. I. Hunter, Jaime.
TJ810 .S65 2021
621.47--dc23

TABLE OF CONTENTS

PREFACE

This book is a culmination of my many years of practice in this field. I attribute the success of this book to my support group. I would like to thank my parents who have showered me with unconditional love and support and my peers and professors for their constant guidance.

Solar power is a form of usable energy that is acquired by converting the energy from sunlight into electric energy. The two main methods used in the conversion of solar energy into electricity are utilizing photovoltaic solar panels and using concentrated solar power. The photovoltaic effect is used by the photovoltaic cells to convert light energy into electric current. Photovoltaic modules use wafer-based crystalline silicon cells and thin-film cells. Lenses, mirrors and tracking systems are used in the concentrated solar power system to focus a large area of sunlight into a small beam. The topics included in this book on solar power are of utmost significance and bound to provide incredible insights to readers. Most of the topics introduced herein cover new techniques and the applications of solar power. As this field is emerging at a rapid pace, the contents of this book will help the readers understand the modern concepts and applications of the subject.

The details of chapters are provided below for a progressive learning:

Chapter - What is Solar Power?

The conversion of energy from sunlight into electricity with the help of photovoltaic cells or concentrated solar power is known as solar power. This is an introductory chapter which will introduce briefly all the important and fundamental concepts of solar power including solar energy, solar power forecasting, etc.

Chapter - Photovoltaics and their Applications

The conversion of light into electricity with the help of semiconducting materials that exhibit the photovoltaic effects is known as photovoltaics. Some of the applications of photovoltaics are for communications, lighting and utility interactive systems. This chapter has been carefully written to provide an easy understanding of the various concepts related to photovoltaics as well as their applications.

Chapter - Solar Cell

The electrical device which is used for the conversion of light energy into electricity by the pho¬tovoltaic effect is known as solar cell. Plasmonic solar cells, thin film solar cells, perovskite solar cells, multi-junction solar cells, dye-sensitized solar cells, organic solar cells, hybrid solar cells and quantum dot solar cells are the various types of solar cells. The topics elaborated in this chapter will help in gaining a better perspective about these types of solar cells.

Chapter - Solar Energy Power Generation

Solar energy can be used for power generation through the conversion of the radiation of the sun into electricity by using solar photovoltaic cells. Some of the components of solar energy power generation are solar panels, solar power batteries and solar inverters. All these components of solar energy power generation have been carefully analyzed in this chapter.

Chapter - Applications of Solar Power

Solar power is used for various purposes such as in solar water heating, solar pool heating, solar cookers, solar street lighting, solar-powered stirling engine, solar furnaces, solar inverters, solar chimneys, etc. The chapter closely examines these key applications of solar power to provide an extensive understanding of the subject.

Chapter - Solar Power Applications in Spacecrafts

Solar power is used to provide power to spacecraft which are operating in the inner solar system. Some of the applications of solar power in spacecraft are in solar thermal rockets, electric solar wind sails and solar electric propulsion. These diverse applications of solar power in spacecraft have been thoroughly discussed in this chapter.

Chapter - Enviornmental Impacts of Solar Power

Solar power has both positive and negative impact upon the environment. The major negative impacts of solar power are habitat loss due to land and water use, and the use of hazardous materials. This chapter closely examines these environmental impacts of solar power as well as the effect of photovoltaic systems and solar farms upon the environment.

Jaime Hunter

What is Solar Power?

CHAPTER 1

The conversion of energy from sunlight into electricity with the help of photovoltaic cells or concentrated solar power is known as solar power. This is an introductory chapter which will introduce briefly all the important and fundamental concepts of solar power including solar energy, solar power forecasting, etc.

SOLAR ENERGY

Solar energy is energy that comes from the sun. Every day the sun radiates, or sends out, an enormous amount of energy. The sun radiates more energy in one second than people have used since the beginning of time.

Where does all this energy come from? It comes from within the sun itself. Like other stars, the sun is a big gas ball made up mostly of hydrogen and helium. The sun generates energy in its core in a process called nuclear fusion. During nuclear fusion, the sun's extremely high pressure and hot temperature cause hydrogen atoms to come apart and their nuclei (the central cores of the atoms) to fuse or combine. Four hydrogen nuclei fuse to become one helium atom. But the helium atom weighs less than the four nuclei that combined to form it. Some matter is lost during nuclear fusion. The lost matter is emitted into space as radiant energy.

It takes millions of years for the energy in the sun's core to make its way to the solar surface, and then just a little over eight minutes to travel the 93 million miles to earth. The solar energy travels to the earth at a speed of 186,000 miles per second, the speed of light.

Only a small portion of the energy radiated by the sun into space strikes the earth, one part in two billion. Yet this amount of energy is enormous. Every day enough energy strikes the United States to supply the nation's energy needs for one and a half years.

Where does all this energy go? About 15 percent of the sun's energy that hits the earth is reflected back into space. Another 30 percent is used to evaporate water, which, lifted into the atmosphere, produce's rain-fall. Solar energy also is absorbed by plants, the land, and the oceans. The rest could be used to supply our energy needs.

People have harnessed solar energy for centuries. As early as the 7th century B.C., people used simple magnifying glasses to concentrate the light of the sun into beams so hot they would cause wood to catch fire. Over 100 years ago in France, a scientist used heat from a solar collector to make steam to drive a steam engine.

In the beginning of this century, scientists and engineers began researching ways to use solar energy in earnest. One important development was a remarkably efficient solar boiler invented by Charles Greeley Abbott, an American astrophysicist, in 1936.

The solar water heater gained popularity at this time in Florida, California, and the Southwest. The industry started in the early 1920s and was in full swing just before World War II. This growth lasted until the mid- 1950s when low-cost natural gas became the primary fuel for heating American homes. The public and world governments remained largely indifferent to the possibilities of solar energy until the oil shortages of the 1970s. Today people use solar energy to heat buildings and water and to generate electricity.

Solar Collectors and Solar Space Heating

Heating with solar energy is not as easy as you might think. Capturing sunlight and putting it to work is difficult because the solar energy that reaches the earth is spread out over a large area. The sun does not deliver that much energy to any one place at any one time. How much solar energy a place receives depends on several conditions. These include the time of day, the season of the year, the latitude of the area, and the clearness or cloudiness of the sky.

A solar collector is one way to collect heat from the sun. A closed car on a sunny day is like a solar collector. As sunlight passes through the car's glass windows, it is absorbed by the seat covers, walls, and floor of the car. The light that is absorbed changes into heat. The car's glass windows let light in, but don't let all the heat out. (This is also why greenhouses work so well and stay warm year-round.)

So, a solar collector does three things:

- It allows sunlight inside the glass (or plastic);
- It absorbs the sunlight and changes it into heat;
- It traps most of the heat inside.

Solar Space Heating

Space heating means heating the space inside a building. Today many homes use solar energy for space heating. There are two general types of solar space heating systems: passive and active. A "hybrid" system is a mixture of the passive and active systems.

Passive Solar Homes

In a passive solar home, the whole house operates as a solar collector. A passive house does not use any special mechanical equipment such as pipes, ducts, fans, or pumps to transfer the heat that the house collects on sunny days. Instead, a passive solar home relies on properly oriented windows. Since the sun shines from the south in North America, passive solar homes are built so that most of the windows face south. They have very few or no windows on the north side.

A passive solar home converts solar energy into heat just as a closed car does. Sunlight passes through a home's windows and is absorbed in the walls and floors.

To control the amount of heat in a passive solar house, the doors and windows are closed or opened to keep heated air in or to let it out. At night, special heavy curtains or shades are pulled over the windows to keep the daytime beat inside the house. In the summer, awnings or roof overhangs help to cool the house by shading the windows from the high summer sun.

Heating a house by warming the walls or floors is more comfortable than heating the air inside a house. It is not so drafty. And passive buildings are quiet, peaceful places to live. A passive solar home can get 50 to 80 percent of the heat it needs from the sun.

Many homeowners install equipment (such as fans to help circulate air) to get more out of their passive solar homes. When special equipment is added to a passive solar home, the result is called a hybrid system.

Active Solar Homes

Unlike a passive solar home, an active solar home uses mechanical equipment, such as pumps and blowers, and an outside source of energy to help heat the house when solar energy is not enough.

Active systems use special solar collectors that look like boxes covered with glass. Dark-colored metal plates inside the boxes absorb the sunlight and change it into heat. (Black absorbs sunlight more than any other color.)

Air or a liquid flows through the collectors and is warmed by this heat. The warmed air or liquid is then distributed to the rest of the house just as it would be with an ordinary furnace system.

Solar collectors are usually placed high on roofs where they can collect the most sunlight. They are also put on the south side of the roof where no tall trees or tall buildings will shade them.

Storing Solar Heat

The challenge confronting any solar heating system - whether passive, active, or hybrid - is heat storage. Solar heating systems must have some way to store the heat that is collected on sunny days to keep people warm at night or on cloudy days.

In passive solar homes, heat is stored by using dense interior materials that retain heat well - masonry, adobe, concrete, stone, or water. These materials absorb surplus heat and radiate it back into the room after dark. Some passive homes have walls up to one foot thick.

In active solar homes, heat may be stored in one of two ways - a large tank may store a hot liquid, or rock bins beneath a house may store hot air.

Houses with active or passive solar heating systems may also have furnaces, wood-burning stoves, or another heat source to provide heat in case there is a long period of cold or cloudy weather. This is called a backup system.

Solar Hot Water Heating

Solar energy is also used to heat water. Water heating is usually the second leading home energy expense, costing the average family over $400, a year. Depending on where you live, and how much hot water your family uses, a solar water heater can pay for itself in as little as five years. A well-maintained system can last 15-20 years, longer than a conventional water heater.

A solar water heater works in the same way as solar space heating. A solar collector is mounted on the roof, or in an area of direct sunlight. It collects sunlight and converts it to heat. When the collector becomes hot enough, a thermostat starts a pump. The pump circulates a fluid, called a heat transfer fluid, through the collector for heating. The heated fluid then goes to a storage tank where it heats water. The hot water may then be piped to a faucet or showerhead. Most solar water heaters that operate in winter use a heat transfer fluid, similar to antifreeze, that will not freeze when the weather turns cold.

Today over 1.5 million homes in the U.S. use solar heaters to heat water for their homes or swimming pools.

Solar Electricity

Besides heating homes and water, solar energy also can be used to produce electricity. Two ways to generate electricity from solar energy are photovoltaics and solar thermal systems.

Photovoltaic Electricity

Photovoltaic comes from the words photo meaning "light" and volt, a measurement of electricity. Sometimes photovoltaic cells are called PV cells or solar cells for short. You are probably already familiar with solar cells. Solar-powered calculators, toys, and telephone call boxes all, use solar cells to convert light into electricity.

A photovoltaic cell is made of two thin slices of silicon sandwiched together and attached to metal wires. The top slice of silicon, called the N-layer, is very thin and has a chemical added to it that provides the layer with an excess of free electrons. The bottom slice, or P-layer, is much thicker and has a chemical added to it so that it has very few free electrons.

When the two layers are placed together, an interesting thing happens-an electric field is produced that prevents the electrons from traveling from the top layer to the bottom layer. This one-way junction with its electric field becomes the central part of the PV cell.

When the PV cell is exposed to sunlight, bundles of light energy known as photons can knock some of the electrons from the bottom P-layer out of their orbits through the electric field set up at the P-N junction and into the N-layer.

The N-layer, with its abundance of electrons, develops an excess of negatively charged electrons. This excess of electrons produces an electric force to push the additional electrons away. These excess electrons are pushed into the metal wire back to the bottom P-layer, which has lost some of its electrons.

This electrical current will continue flowing as long as radiant energy in the form of light strikes the cell and the pathway, or circuit, remains closed.

Current PV cell technology is not very efficient. Today's PV cells convert only about 10 to 14 percent of the radiant energy into electrical energy. Fossil fuel plants, on the other hand, convert from 30-40 percent of their fuel's chemical energy into electrical energy. The cost per kilowatt-hour to produce electricity from PV cells is presently three to four times as expensive as from conventional sources. However, PV cells make sense for many uses today, such as providing power in remote areas or other areas where electricity is difficult to provide. Scientists are researching ways to improve PV cell technology to make it more competitive with conventional sources.

Solar Thermal Electricity

Like solar cells, solar thermal systems use solar energy to make electricity. But as the name suggests, solar thermal systems use the sun's heat to do it.

Most solar thermal systems use solar collectors with mirrored surfaces to concentrate sunlight onto a receiver that heats a liquid. The super-heated liquid is used to make steam that drives a turbine to produce electricity in the same way that coal, oil, or nuclear power plants do.

Solar thermal systems may be one of three types: central receiver, dish, or trough. A central receiver system uses large mirrors on top of a high tower to reflect sunlight onto a receiver. This system has been dubbed a "solar power tower." Another system uses a dish-shaped solar collector to collect sunlight. This system resembles a television satellite dish. A third system uses mirrored troughs to collect sunlight. Until recently, trough systems seemed the most promising.

The world's first solar electric plant used mirrored troughs. LUZ, as the plant was called, was perfectly situated in the sunny Mojave desert of California. LUZ was the only solar plant to generate electricity economically. Dollar for dollar, it had always been cheaper to use conventional sources of energy (coal, oil, nuclear) to generate electricity. But the LUZ solar plant turned that around, producing electricity as cheaply as many new coal plants, and with no hidden pollution costs. The future looked bright for this pioneering solar plant and then the dream cracked. LUZ closed its doors at the end of 1992 because of a drop in oil prices and an over-budget construction project at LUZ's home-base.

LUZ may be gone, but most solar energy engineers believe solar power towers will be ready to take the place of trough systems very soon.

Solar Energy and the Environment

In the 1970s, the push for renewable energy sources was driven by oil shortages and price increases. Today, the push for renewable energy sources is driven by a renewed concern for the environment.

Solar energy is the prototype of an environmentally friendly energy source. It consumes none of our precious energy resources, makes no contribution to air, water, or noise pollution, does not pose a health hazard, and contributes no harmful waste products to the environment.

There are other advantages too. Solar energy cannot be embargoed or controlled by any one nation. And it will not run out until the sun goes out.

Advantages of Solar Energy

Renewable Energy Source

Among all the benefits of solar panels, the most important thing is that solar energy is a truly renewable energy source. It can be harnessed in all areas of the world and is available every day. We cannot run out of solar energy, unlike some of the other sources of energy. Solar energy will be accessible as long as we have the sun, therefore sunlight will be available to us for at least 5 billion years when according to scientists the sun is going to die.

Reduces Electricity Bills

Since you will be meeting some of your energy needs with the electricity your solar system has generated, your energy bills will drop. How much you save on your bill will be dependent on the size of the solar system and your electricity or heat usage. Moreover, not only will you be saving on the electricity bill, there is also a possibility to receive payments for the surplus energy that you export back to the grid. if you generate more electricity than you use (considering that your solar panel system is connected to the grid).

Diverse Applications

Solar energy can be used for diverse purposes. You can generate electricity (photovoltaics) or heat (solar thermal). Solar energy can be used to produce electricity in areas without access to the energy grid, to distill water in regions with limited clean water supplies and to power satellites in space. Solar energy can also be integrated into the materials used for buildings. Not long ago Sharp introduced transparent solar energy windows.

Low Maintenance Costs

Solar energy systems generally don't require a lot of maintenance. You only need to keep them relatively clean, so cleaning them a couple of times per year will do the job. If in doubt, you can always rely on specialised cleaning companies, which offer this service from around £25-£35. Most reliable solar panel manufacturers offer 20-25 years warranty. Also, as there are no moving parts, there is no wear and tear. The inverter is usually the only part that needs to changed after 5-10 years because it is continuously working to convert solar energy into electricity and heat (solar PV vs. solar thermal). Apart from the inverter, the cables also need maintenance to ensure your solar power system runs at maximum efficiency. So, after covering the initial cost of the solar system, you can expect very little spending on maintenance and repair work.

Technology Development

Technology in the solar power industry is constantly advancing and improvements will intensify in the future. Innovations in quantum physics and nanotechnology can potentially increase the effectiveness of solar panels and double, or even triple, the electrical input of the solar power systems.

Disadvantages of Solar Energy

Cost

The initial cost of purchasing a solar system is fairly high. This includes paying for solar panels, inverter, batteries, wiring, and for the installation. Nevertheless, solar technologies are constantly developing, so it is safe to assume that prices will go down in the future.

Weather Dependent

Although solar energy can still be collected during cloudy and rainy days, the efficiency of the solar system drops. Solar panels are dependent on sunlight to effectively gather solar energy. Therefore, a few cloudy, rainy days can have a noticeable effect on the energy system. You should also take into account that solar energy cannot be collected during the night. On the other hand, if you also require your water heating solution to work at night or during wintertime, thermodynamic panels are an alternative to consider.

Solar Energy Storage is Expensive

Solar energy has to be used right away, or it can be stored in large batteries. These batteries, used in off-the-grid solar systems, can be charged during the day so that the energy is used at night. This is a good solution for using solar energy all day long but it is also quite expensive. In most cases, it is smarter to just use solar energy during the day and take energy from the grid during the night (you can only do this if your system is connected to the grid). Luckily your energy demand is usually higher during the day so you can meet most of it with solar energy.

Uses a Lot of Space

The more electricity you want to produce, the more solar panels you will need, as you want to collect as much sunlight as possible. Solar PV panels require a lot of space and some roofs are not big enough to fit the number of solar panels that you would like to have. An alternative is to install some of the panels in your yard but they need to have access to sunlight. If you don't have the space for all the panels that you wanted, you can opt for installing fewer to still satisfy some of your energy needs.

Associated with Pollution

Although pollution related to solar energy systems is far less compared to other sources of energy, solar energy can be associated with pollution. Transportation and installation of solar systems have been associated with the emission of greenhouse gases. There are also some toxic materials and hazardous products used during the manufacturing process of solar photovoltaic systems, which can indirectly affect the environment. Nevertheless, solar energy pollutes far less than other alternative energy sources.

SOLAR POWER

Solar power is obtained by tapping the sun's energy and converting it into electricity using solar energy technologies like solar panels or concentrated power. Solar power is the most renewable form of energy existing today. Solar power is classified as renewable energy source as it's eternal on a human time scale. It does not produce harmful emissions, and most importantly, it's absolutely free. On the contrary, fossils fuels like oil, gas, and coal are non-renewable energy resources since they are in limited supply. Simply stated, we will run out of fossils fuels some day.

Working of Solar Power

The sun rays contain vast amounts of energy. Naturally, when the sun rays strike an object, the energy is immediately converted into heat. Think of the warmth you perceive when lying in the sun. However, when the sun rays strike certain specially designed materials, it's converted into electrical current, which can be tapped for electricity. These specially designed devices capable of converting the sun's energy into electricity are called solar panels or photovoltaic panels.

A solar panel is made up of multiple solar cells or photovoltaic cells, which are chiefly responsible for converting the sun's energy into electricity. Photovoltaic is the scientific process whereby solar energy is converted into electricity. Each solar panel or photovoltaic panel comes with a given number of silicon cells. These cells react with photons existing in the sun rays to generate direct current (DC).

The direct current is then channeled to an inverted where it's converted into Alternating Current (AC). The Alternating current is transmitted to your home or office distribution network and is now able to power your home or office appliances. When you produce more electricity than you need, the surplus is fed back into the national power grid, which can earn you some money.

Most Popular Methods to Collect Solar Energy

There are few methods used to convert the sun's energy into electricity, but the most common ones are solar panels and solar thermal systems.

Solar Thermal Systems

Solar thermal systems are primarily applied to residential buildings where there is greater need to lower monthly energy bills. However, this technology can also be utilized in commercial buildings. The idea behind solar thermal systems is to tap the sun's energy, convert it into heat and channel it into your home or business heating system as hot water or space heating. A wide variety of approaches can be applied here including active and passive systems.

A typical example of an active system is the solar hot water heater. Solar hot water heaters come in huge range of designs and include a collector immersion heater and a storage tank. The solar collector is usually black in color to absorb maximum energy from the sun. The collector contains a transfer liquid, which is normally a blend of water and glycol, commonly known as antifreeze. This combination helps to prevent freezing of water during winter.

The heated water in the collector is then channeled to a heat exchanger situated in the storage tank in your home. The heat coming from the exchanger will heat up the water contained in the storage tank. After the liquid emits its heat, the water is allowed to flow back to the collector for further heating. The function of the controller is to make sure the liquid circulates to the collector when enough heat is available.

Passive systems involve innovative engineering designs that culminate to a building that automatically stores up and uses solar energy. A typical example of passive system used to generate solar energy is the greenhouse. A greenhouse gathers solar energy on a sunny day and uses it to keep your house warm during the night.

How the Greenhouse Principle Works?

The sun rays are trapped by the greenhouse action of strategically cited (north facing) glass areas that are entirely exposed to the sun. The efficiency of the whole process depends on window orientation, frames, shades and glazing type. The trapped heat is later absorbed and stored by high thermal materials in the house. The heat is released during the night to cater for the lost heat in order to lower maintain the required temperature in the house.

Passive solar heating is utilized along with passive shading, which enables maximum absorption of solar energy during winter and mitigates overheating during summer. This heat balance is attained by properly citing the areas of the glass (northern orientation) including properly designed eave overhangs. Re-radiated heat is well circulated to where it's needed via an effective air flow design and convection.

The most effective form of heat circulation is direct re-radiation. However, heat is also conducted via building materials and circulated by air movement. Designing the floor should also be taken into account when looking to apply the greenhouse principle. The most essential rooms such as living rooms should be designed facing north to ensure they get the best winter solar access.

Heat loss is substantially reduced with the right window treatments and impeccably insulated attic, walls, and raised floor. The thermal storage system must also be properly insulated to achieve high levels of effectiveness.

Quality windows and doors, airlocks, airtight construction detailing, and draught sealing substantially reduce air infiltration. The overall shape of the house and room layout should be designed appropriately to reduce heat loss, which occurs from every part of the building, but rapidly through the roof. In cold climates, it's more efficient to have compact shapes that substantially reduce roof and external wall area. When the climate becomes warm, more external wall is essential to enable incredible cross-ventilation.

Solar Photovoltaics or Solar Panels

Photovoltaic involves direct conversion of sun rays into electricity at the atom level. Specific materials manifest a characteristic called the photoelectric effect that triggers them to absorb photons contained in the sun rays and emit electrons. These free electrons are later captured resulting in an electric current that can be utilized as electricity.

How Solar Photovoltaic Works?

Photovoltaic cells or solar cells are made of semiconductor material known as silicon. It all starts by a thin semiconductor wafer contained in the solar cells. The semiconductor wafer is carefully treated to create an electric field; positive field on one side and negative field on the other. When the sun rays hit the solar cell, electrons are knocked off the atoms in the semiconductor material. An electric circuit is formed when electrical conductors are tethered to the positive and negative sides, allowing the electrons to be harnessed as electrical current (electricity). This electricity can be used to power homes or businesses.

When a given number of solar cells are electrically tethered to each other, and set up on a support structure or frame, it's referred to as a photovoltaic module. Modules are specially developed to provide electricity at a predetermined voltage, like the common 12-volt system. The strength of the voltage produced depends upon the intensity of sun rays striking the surface of the solar cell.

Numerous modules can be joined to form an array. Generally, if the area of the array is larger, more electric power will be generated. Photovoltaic cells produce direct current (DC). The direct current is transmitted to a converter that turns it into Alternating Current (AC) that is able to power home and office appliances.

SOLAR POWER FORECASTING

Solar power forecasting involves knowledge of the Sun´s path, the atmosphere's condition, the scattering processes and the characteristics of a solar energy plant which utilizes the Sun's energy to create solar power. Solar photovoltaic systems transform solar energy into electric power. The power output depends on the incoming radiation and on the solar panel characteristics. Photovoltaic power production is increasing nowadays. Forecast information is essential for an efficient use, the management of the electricity grid and for solar energy trading. Common solar forecasting method include stochastic learning method, local and remote sensing method, and hybrid method.

Generation Forecasting

The energy generation forecasting problem is closely linked to the problem of *weather variables forecasting*. Indeed, this problem is usually split into two parts, on one hand focusing on the forecasting of solar PV or any other meteorological variable and on the other hand estimating the amount of energy that a concrete power plant will produce with the estimated meteorological resource. In general, the way to deal with this difficult problem is usually related to the spatial and temporal scales we are interested in, which yields to different approaches that can be found in the literature. In this sense, it is useful to classify these techniques depending on the forecasting horizon, so it is possible to distinguish between *now-casting* (forecasting 3–4 hours ahead), *short-term forecasting* (up to 7 days ahead) and *long-term forecasting* (months, years) Solar radiation closely follows the physical and biological development of the earth. Its spatial and sequential heterogeneity powerfully influence the forcing of environmental and hydrological organisms by manipulating air temperature, soil moisture and vapor transpiration, snow cover and lots of photochemical procedures. Therefore, solar radiation drives place efficiency and plant life allotment, organism a key feature in undeveloped and forestry sciences that be obliged to be known precisely. The quantity of solar radiation obtainable at the earth' surface is at the outset controlled at worldwide balance, organism above all precious by the Sun Earth geometry and the atmosphere. On the other hand, a complete explanation of its freedom time unpredictability require deliberation of limited procedure which frequently turn out to be also applicable, as is the casing in mountainous region. Predominantly, limited territory adjust the inward bound solar radiation by shadow casts, slope of elevation, surface gradient and compass reading, as a result, precise spatial model of inward bound solar radiation be supposed to regard as the pressure of the terrain surface. In the final time, more than a few events to consist of the confined terrain special effects in the solar radiation countryside have been projected, such as the use of Geographical Information Systems (GIS), artificial intelligence or post dispensation of satellite stand technique. Solar radiation can be also evaluated using numerical weather forecast (NWP) models. Nevertheless, the space and time balance determined with them and the incomplete computational ability frequently avoid the deliberation of terrain connected property.

Otherwise, exclamation technique agree to us to acquire spatially persistent database from data evidence at inaccessible station greater than wide region. Even though their dependability is powerfully needy on the opening coldness between position, they eventually rely on experiential statistics, which have a superior precision than extra method. Therefore, while an adequate footage spatial thickness is accessible, disturbance method are preferred. Conventionally, solar radiation has not been as densely example as additional variables as temperature or rainfall, therefore the ease of use of capacity is frequently in short supply. Though, the number of experimental system which record solar radiation has developed and interruption has been converted into an appropriate technique for solar radiation evaluation. Nevertheless, radiometric stations are frequently come together approximately farmland or occupied region, typically during basin and plane area, while mountains at rest require enough footage thickness. This truth is particularly applicable afford the tall spatial unpredictability of solar radiation in these province. As an outcome, particular interruption method that tolerate include outdoor foundation should be used to make clear this extra spatial unpredictability. Several diverse spatial interruption techniques can be established. On the other hand, data ease of use in mountainous region is often extremely restricted. As a result, it is hard to construct a precise solar radiation climatology in hilly area to be used in environmental science, climate change.

Solar radiation is a hardly illustration changeable with reverence to supplementary ecological variables such as temperature or precipitation, in fraction payable to the high maintenance price of the necessary radiometric sensors. It is extremely sensitive to ecological feature on or after local to limited balance. Predominantly, terrain surface confronts the conventional interruption method while forecast through far above the ground spatial decision are required, particularly due to the lack of measurement stations in mountainous areas. Geo-statistics front a stochastic move toward to resolve the spatial forecast difficulty that stop dependence on before imagine deterministic models and permit us to consist of the consequence of outside in sequence foundation stand on investigation data-sets.

Nowcasting

The term “Nowcasting” in the context of solar power forecasting, generally refers to same spatial and temporal scales as meteorological Nowcasting, which focuses on forecast horizons from a few minutes ahead, out to 4-6 hours ahead. In general, the ‘Nowcasting’ forecast horizon refers to those not well-served by global numerical weather prediction models, which produce outputs at up to hourly resolutions and only update every 6 hours. Solar power nowcasting then refers to the prediction of solar power output (or energy generation) over time horizons of tens to hundreds of minutes ahead of time with up to 90% predictability. It has historically been very important for electrical grid operators in order to guarantee the matching of supply and demand on energy markets. Such solar power nowcasting services are usually related to temporal resolutions of 5 to 15 minutes, with updates as frequent as every 5 minutes. The regular updates and relatively high resolutions required from these methods require automatic weather data acquisition and processing techniques, which are chiefly accomplished by two primary means:

1. Statistical techniques: These are usually based on time series processing of measurement data, including meteorological observations and power output measurements from a solar power facility. What then follows is the creation of a training dataset to tune the parameters of a model, before evaluation of model performance against a separate testing dataset. This class of techniques includes the use of any kind of statistical approach, such as autoregressive moving averages (ARMA, ARIMA, etc), as well as machine learning techniques such as neural networks, support vector machines (etc). These approaches are usually benchmarked to a persistence approach in order to evaluate their improvements. This persistence approach just assumes that any variable at time step t is the value it took in a previous time.
2. Satellite based methods: These methods leverage the several geostationary Earth observing

weather satellites (such as Meteosat Second Generation (MSG) fleet) to detect, characterise, track and predict the future locations of cloud cover. These satellites make it possible to generate solar power forecasts over broad regions through the application of image processing and forecasting algorithms. Key forecasting algorithms include cloud motion vectors (CMVs). Relevant methods for applying physical models based on satellite image processing techniques provide an estimation of future atmospheric values can be found in Alvarez et al.

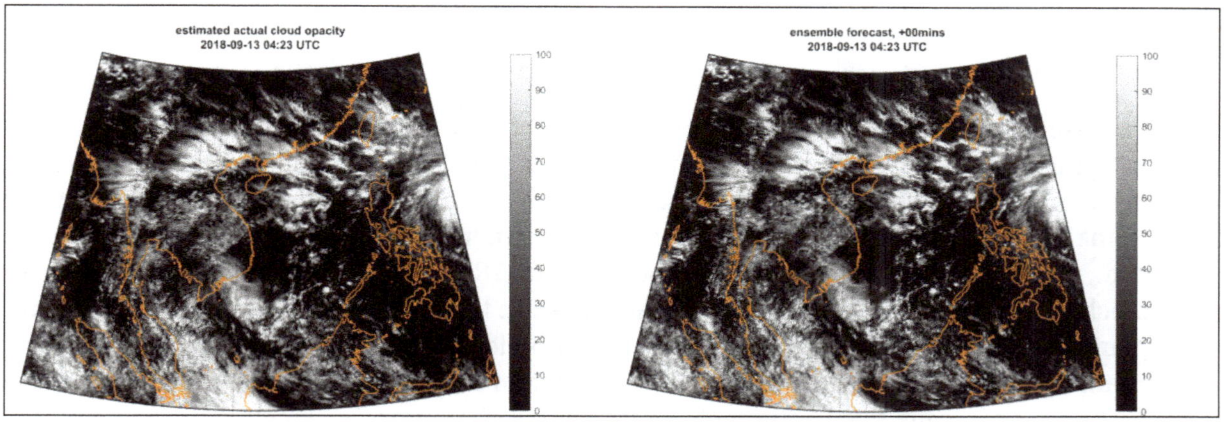

An example of satellite-based cloud cover nowcasting, which is used to generate predication of solar power outputs.

Short-term Solar Power Forecasting

Short-term forecasting provides predictions up to 7 days ahead. This kind of forecast is also valuable for grid operators in order to make decisions of grid operation, as well as, for electric market operators. Under this perspective, the meteorological resources are estimated at a different temporal and spatial resolution. This implies that meteorological variables and phenomena are looked from a more general perspective, not as local as nowcasting services do. In this sense, most of the approaches make use of different numerical weather prediction models (NWP) that provide an initial estimation of weather variables. Currently, several models are available for this purpose, such as Global Forecast System (GFS) or data provided by the European Center for Medium Range Weather Forecasting (ECMWF). These two models are considered the state of the art of global forecast models, which provide meteorological forecasts all over the world. In order to increase spatial and temporal resolution of these models, other models have been developed which are generally called mesoscale models. Among others, HIRLAM, WRF or MM5 are the most representative of these models since they are widely used by different communities. To run these models a wide expertise is needed in order to obtain accurate results, due to the wide variety of parameters that can be configured in the models. In addition, sophisticated techniques such as data assimilation might be used in order to produce more realistic simulations. Finally, some communities argue for the use of post-processing techniques, once the models' output is obtained, in order to obtain a probabilistic point of view of the accuracy of the output. This is usually done with ensemble techniques that mix different outputs of different models perturbed in strategic meteorological values and finally provide a better estimate of those variables and a degree of uncertainty, like in the model proposed by Bacher et al.

Long-term Solar Power Forecasting

Long-term forecasting usually refers to forecasting of the annual or monthly available resource. This is useful for energy producers and to negotiate contracts with financial entities or utilities that distribute the generated energy. In general, these long-term forecasting is usually done at a lower scale than any of the two previous approaches. Hence, most of these models are run with meso-scale models fed with reanalysis data as input and whose output is postprocessed with statistical approaches based on measured data.

Energetic Models

This step is usually done with statistical approaches that try to correlate the amount of available resource with the metered power output. The main advantage of these methods is that the meteorological prediction error, which is the main component of the global error, might be reduced taking into account the uncertainty of the prediction. As it was mentioned before and detailed in Heinemann et al., these statistical approaches comprises from ARMA models, neural networks, support vector machines, etc. On the other hand, there also exist theoretical models that describe how a power plant converts the meteorological resource into electric energy, as described in Alonso et al. The main advantage of this type of models is that when they are fitted, they are really accurate, although they are too sensitive to the meteorological prediction error, which is usually amplified by these models. Hybrid models, finally, are a combination of these two models and they seem to be a promising approach that can outperform each of them individually.

References

- Solar, texts, essence: lsa.colorado.edu, Retrieved 8 May, 2019
- Sanjari, M.J.; Gooi, H.B. (2016). "Probabilistic Forecast of PV Power Generation based on Higher-order Markov Chain". IEEE Transactions on Power Systems. Doi:10.1109/TPWRS.2016.2616902
- 5-advantages-and-5-disadvantages-of-solar-energy: greenmatch.co.uk, Retrieved 9 June, 2019
- "Solar Energy Forecasting and Resource Assessment - 1st Edition". www.elsevier.com. Retrieved 2019-05-08
- Whatsolarpower: conserve-energy-future.com, Retrieved 10 July, 2019
- Espino, I., Hernández, M.. Nowcasting of wind speed using support vector regression. Experiments with Time Series from Gran Canaria. Renewable Energy and Power Quality Journal, ISSN 2172-038X, N9, 12 May 2011
- How-is-solar-energy-stored: solar-energy-at-home.com, Retrieved 11 August, 2019

Photovoltaics and their Applications

2 CHAPTER

The conversion of light into electricity with the help of semiconducting materials that exhibit the photovoltaic effects is known as photovoltaics. Some of the applications of photovoltaics are for communications, lighting and utility interactive systems. This chapter has been carefully written to provide an easy understanding of the various concepts related to photovoltaics as well as their applications.

PHOTOVOLTAICS

Photovoltaic (PV) devices generate electricity directly from sunlight via an electronic process that occurs naturally in certain types of material, called semiconductors. Electrons in these materials are freed by solar energy and can be induced to travel through an electrical circuit, powering electrical devices or sending electricity to the grid.

PV devices can be used to power anything from small electronics such as calculators and road signs up to homes and large commercial businesses.

The PV effect was observed as early as 1839 by Alexandre Edmund Becquerel, and was the subject of scientific inquiry through the early twentieth century. In 1954, Bell Labs in the U.S. introduced the first solar PV device that produced a useable amount of electricity, and by 1958, solar cells were being used in a variety of small-scale scientific and commercial applications.

The energy crisis of the 1970s saw the beginning of major interest in using solar cells to produce electricity in homes and businesses, but prohibitive prices (nearly 30 times higher than the current price) made large-scale applications impractical.

Industry developments and research in the following years made PV devices more feasible and a cycle of increasing production and decreasing costs began which continues even today.

How does PV Technology Work?

Photons strike and ionize semiconductor material on the solar panel, causing outer electrons to break free of their atomic bonds. Due to the semiconductor structure, the electrons are forced in one direction creating a flow of electrical current. Solar cells are not 100% efficient in crystalline silicon solar cells, in part because only certain light within the spectrum can be absorbed. Some of the light spectrum is reflected, some is too weak to create electricity (infrared) and some (ultraviolet) creates heat energy instead of electricity.

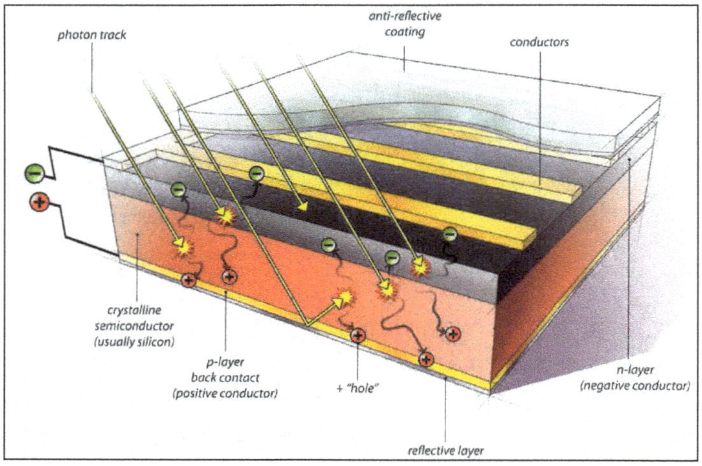

Diagram of a typical crystalline silicon solar cell. To make this type of cell, wafers of high-purity silicon are "doped" with various impurities and fused together. The resulting structure creates a pathway for electrical current within and between the solar cells.

Other Types of Photovoltaic Technology

In addition to crystalline silicon (c-Si), there are two other main types of PV technology:

- Thin-film PV is a fast-growing but small part of the commercial solar market. Many thin-film firms are start-ups developing experimental technologies. They are generally less efficient – but often cheaper – than c-Si modules.
- In the United States, concentrating PV arrays are found primarily in the desert Southwest. They use lenses and mirrors to reflect concentrated solar energy onto high-efficiency cells. They require direct sunlight and tracking systems to be most effective.
- Building-integrated photovoltaics serve as both the outer layer of a structure and generate electricity for on-site use or export to the grid. BIPV systems can provide savings in materials and electricity costs, reduce pollution, and add to the architectural appeal of a building.

Costs of Solar Photovoltaics

Rapidly falling prices have made solar more affordable than ever. The average price of a completed PV system has dropped by 59 percent over the last decade.

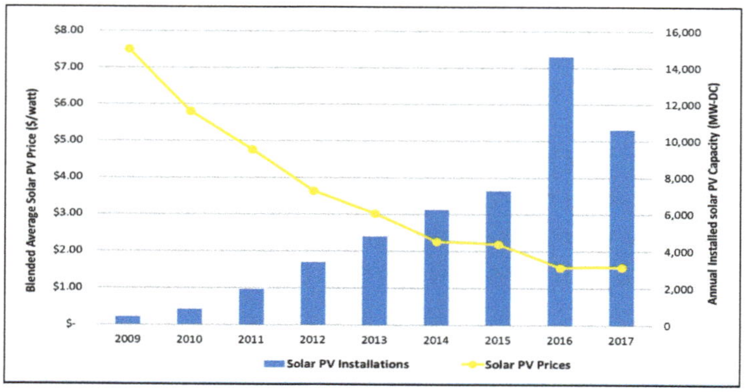

Modern Photovoltaics

The cost of PV has dropped dramatically as the industry has scaled up manufacturing and incrementally improved the technology with new materials. Installation costs have come down too with more experienced and trained installers. Globally, the U.S. has the third largest market for PV installations, and is continuing to rapidly grow.

Most modern solar cells are made from either crystalline silicon or thin-film semiconductor material. Silicon cells are more efficient at converting sunlight to electricity, but generally have higher manufacturing costs. Thin-film materials typically have lower efficiencies, but can be simpler and less costly to manufacture. A specialized category of solar cells - called multi-junction or tandem cells - are used in applications requiring very low weight and very high efficiencies, such as satellites and military applications. All types of PV systems are widely used today in a variety of applications.

PHOTOVOLTAIC EFFECT

The photovoltaic effect is a process that generates voltage or electric current in a photovoltaic cell when it is exposed to sunlight.

The key property of a photovoltaic material is to convert light energy to electric current. This conversion takes place due to the photovoltaic effect - a physical phenomenon in a semiconductor.

Semiconductors are a special class of materials, whose conductance is not permanent, but rather depends on the energy available to activate electrons in the crystal lattice. Crystalline silicon is a semiconductor material widely used in photovoltaics. It becomes conductive when energy of the photons absorbed by the crystal surface is sufficient to raise the electron state from the valence band to the conduction band. This required amount of energy to excite an electron is defined as band gap. Band gap is an intrinsic property of semiconductors and eventually has a direct influence on the photovoltaic cell voltage. The following schematic provides a demonstration of the band gap concept.

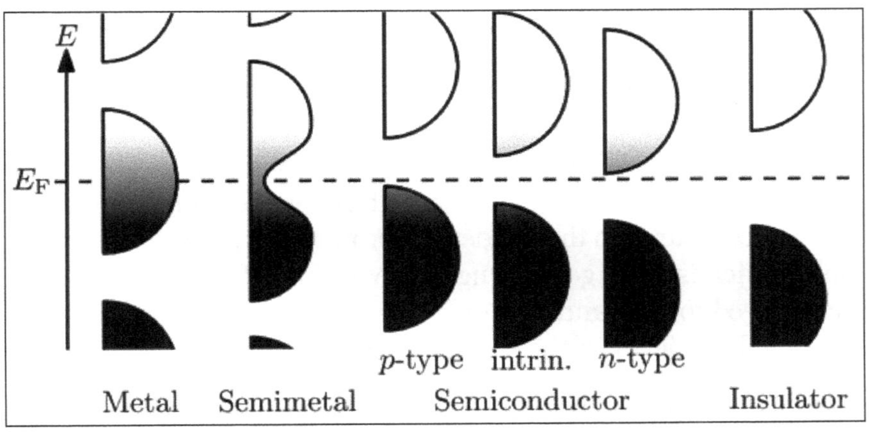

Schematic illustration of the band gaps in various materials. The vertical axis is the electron energy, and EF is the position of the Fermi level.

In this picture, we can visualize the difference between different classes of materials: conductors (metals), semiconductors, and insulators. The valence and conduction bands in a metal overlap, so it does not take any significant energy to free the electrons. They are available for conduction as soon as the potential gradient is provided. In insulators, the gap between the valence and conduction bands is very large, so it requires so much energy to free the electrons that it can damage the material itself. For semiconductors, the situation is somewhere in between. The band gap is big enough to prevent spontaneous conduction and to provide separation of charges, and small enough to be matched by photon energy.

Some of those materials are more suitable for photovoltaic applications than others. How suitable the materials are for photovoltaic applications would be determined by how close the photon energy is to the band gap of the material, and if energy of photon is sufficient to cover the band gap.

Table: The band gap energies of several different materials.

Material		Band Gap (eV)
Si	Silicon	1.11
Ge	Germanium	0.67
CdTe	Cadmium telluride	1.5
Cu_2O	Copper oxide	2.1
Cu_3N	Copper nitride	1.75
GaP	Gallium phosphide	2.26
GaAs	Gallium arsenide	1.43
PbS	Lead sulfide	0.37
Si_3N_4	Silicon nitride	5
C	Diamond	5.5
SiO_2	Silica	9

Band gaps of different materials.

For example, if we have a photon with energy of 2 eV hitting silicon surface, 1.1 eV of that energy will be used to move an electron to the conduction band; the rest of the energy (0.9 eV) will be dissipated as heat. However, if a material with a greater band gap is used, for instance copper oxide, 2 eV is not enough to free the electron. We would need a higher energy photon there.

Based on these reflections, we see that low band gap materials (such as germanium), can be used to capture low energy photons (like those in the red and IR parts of spectrum), and high band gap materials can be used for using high energy photons. Combination of different materials in one system allows for more efficient use of available radiation. At the same time, if the band gap of the PV material is too small compared to the incident photon energy, significant amount of energy will be converted to heat, which is not a good thing for PV cell itself. No matter how much higher the photon energy is compared to the band gap, only one electron can be freed by one photon. This is the reason for the limited efficiency of the photovoltaic cells.

The data in figure show how the maximum efficiency of a solar cell depends on the band gap. If the band gap is too high, most photons will not cause photovoltaic effect; if it is too low, most photons will have more energy than necessary to excite electrons across the band gap, and the rest of energy will be wasted. The semiconductors commonly used in commercial solar cells have band gaps near the peak of this curve, for example silicon (1.1eV) or CdTe (1.5eV). The Shockley–Queisser limit (33.7%) defined at the peak of the curve has been exceeded experimentally by combining materials with different band gap energies into tandem solar cells.

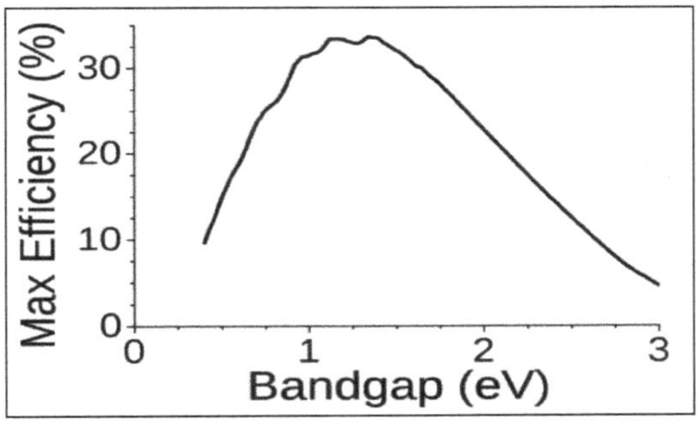

The prediction of the solar cell maximum efficiency as a function material band gap.

PHOTOVOLTAIC THERMAL HYBRID SOLAR COLLECTOR

Photovoltaic thermal hybrid solar collectors, also known as hybrid PV/T (PVT) or solar cogeneration systems, are power generation technologies that convert solar radiation into usable thermal and electrical energy. Such systems combine a solar cell, which converts sunlight into electricity, with a solar thermal collector, which captures the remaining energy and removes waste heat from the PV module. These technologies can be more energy efficient overall than solar photovoltaic (PV) or solar thermal alone.

Significant research has gone into developing a diverse range of PV/T technologies since the 1970s. While many collector types have been developed for small-scale residential or specific process use, powerful and highly-efficient systems for industrial and utility-scale applications are also beginning to enter the market. Collectors co-generating up to 250kW electricity plus 400kW heat, and operating with 80% conversion efficiency, are commercially available as of 2017.

PV/T System Engineering

Photovoltaic cells suffer from a drop in efficiency with the rise in temperature due to increased resistance. PV/T systems are engineered to actively capture and carry heat away from the PV cells, thereby cooling the cells and thus improving their efficiency by lowering resistance. Cell lifetimes also improve at lower temperatures, and the thermal capture can minimize the urban heat island contribution from the generator.

This is an effective method to maximize total system efficiency and reliability, but causes the thermal component to under-perform as compared to that achievable with a pure solar thermal collector. That is to say, the maximum operating temperatures for most PV/T system are limited to less than the maximum cell temperature (typically below 100 °C). Nevertheless, two or more units of heat energy are still generated for each unit of electrical energy, depending on cell efficiency and system design.

System Types

A number of PV/T collectors in different categories are commercially available and can be divided into the following categories:

- PV/T liquid collector.
- PV/T air collector.
- PV/Ta Liquid and air collector.
- PV/T concentrator (CPVT).

PV/T Liquid Collector

The basic water-cooled design uses a channel to direct fluid flow using piping of various materials or plates attached to the back of a PV module. The fluid flow arrangement through the cooling element will determine which systems the panels are most suited to.

In a standard fluid-based system, a working fluid, typically water, glycol or mineral oil is then piped through these pipes or plate chillers. The heat from the PV cells is conducted through the metal and absorbed by the working fluid (presuming that the working fluid is cooler than the operating temperature of the cells). In closed-loop systems this heat is either exhausted (to cool it), or transferred at a heat exchanger, where it flows to its application. In open-loop systems, this heat is used, or exhausted before the fluid returns to the PV cells. It is also possible to disperse nanoparticles in the liquid to create a liquid filter for PV/T applications. The basic advantage of this type of split configuration is that the thermal collector and the photovoltaic collector can operate at different temperatures.

PV/T Air Collector

The basic air-cooled design uses a hollow, conductive metal housing to mount the photo-voltaic (PV) panels. Heat is radiated from the panels into the enclosed space, where the air is either circulated into a building HVAC system to recapture heat energy, or rises and is vented from the top of the structure.

While energy transfer to air is not as efficient as a liquid collector, the infrastructure required has lower cost and complexity; basically a shallow metal box. Placement of PV panels can be vertical or angled.

PV/T Concentrator (CPVT)

A concentrator system has the advantage to reduce the amount of photovoltaic (PV) cells needed, such that somewhat more expensive and efficient multi-junction photovoltaic cells can be used that will maximize the ratio of produced high-value electrical power versus lower-value thermal power. Concentrator systems also often require reliable control systems to accurately track the sun and to protect the PV cells from damaging over-temperature conditions. Under ideal conditions, about 75% of the sun's power directly incident upon such systems can be gathered as electricity and heat. For more details, see the discussion of CPVT within the article for concentrated photovoltaics.

A limitation of high-concentrator (i.e. HCPV and HCPVT) systems is that they maintain their long-term advantages over conventional c-Si/mc-Si collectors only in regions that remain consistently free of atmospheric aerosol contaminants (e.g. light clouds, smog, etc.). Power production is rapidly degraded because 1) radiation is reflected and scattered outside of the small (often less than 1°-2°) acceptance angle of the collection optics, and 2) absorption of specific components of the solar spectrum causes one or more series junctions within the MJ cells to under-perform. The short-term impacts of such power generation irregularities can be reduced to some degree with inclusion of electrical and thermal storage in the system.

APPLICATIONS OF PHOTOVOLTAICS

With the ever increasing demand for energy, the search for alternative energy sources has increased. The worldwide use of fossil fuels has led to the critical situation of global warming, significantly effecting our health, environment and climate. Extensive emphasis have been put on the implementation of renewable energy sources. Solar energy is by far the most abundant form of renewable energy and has the potential to partially replace fossil fuels. The amount of solar radiation striking our earth's surface is about ten thousand times higher than the current global electrical energy consumption. Photovoltaic (PV) cells is one of the way to harness solar energy. PV cells convert sunlight directly to electricity and can be influential in meeting the world's energy demand. PV systems are being used in a variety of applications. These applications may be grouped into two categories: "utility interactive systems" and "stand-alone systems".

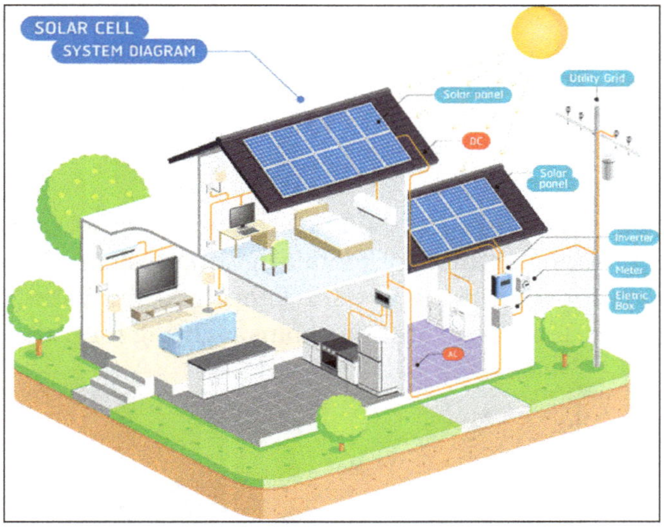

Schematics of utility interactive applications.

Utility Interactive Applications

In utility interactive (or grid-connected) PV systems, PV modules are connected to inverters that convert the DC produced by the PV modules to AC. This electricity can then power household appliances or can be sold directly to the grid. As a building receives this energy, it is distributed to appliances and lighting, or other devices where needed. Since PV systems are restricted to function only exposed to the sun, a backup system is frequently required to ensure continuous supply of electricity irrespective of the weather conditions. These systems are most commonly used in houses or commercial buildings to offset electricity cost. A well designed PV system with a proper storage facility can be an attractive prospect for displacing power during the peak hours.

The schematic above shows a typical photovoltaic panel system, on the roof of a house. This system allows the owner to sell generated electricity to the National Grid, effectively lowering the house owners' electricity bills.

PV panels on the roof of a house.

Stand-alone Systems

Stand-alone systems directly use the generated produced electricity. Stand-alone systems do not rely on utility/grid connections. When the requirement arises during night time or poor sunlight, a battery storage system is used. In some situations, stand-alone systems use conventional generators as backup systems. There are numerous applications of stand-alone PV systems.

Lighting

With the invention of LED (light emitting diode) technology as low power lighting sources, PV systems find an ideal application in remote or mobile lighting systems. PV systems combined with battery storage facilties are mostly used to provide lighting for billboards, highway in formation signs, public-use facilities, parking lots, vacation cabins, lighting for trains.

A street light powered by PV.

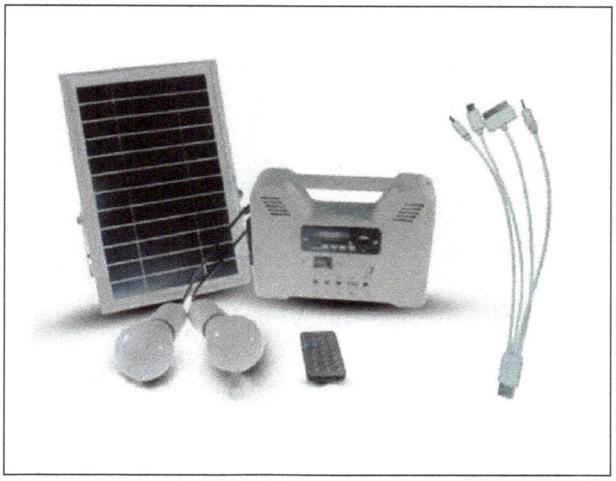

Portable lighting system along with mobile charging facility.

Example of a bill board powered by solar panels.

Communications

Signals required by communication systems need amplification after intervals. Various relay towers are stationed to boost High grounds are mostly favoured as the sites for repeater stations. Th far from power lines. To reduce the difficulty and systems are being installed as a viable alternative.

Repeater at an elevated location can be powered by PV.

Satellites used for communication are powered by PV.

Electricity for Remote Areas

Some areas are quite far from the distribution network to establish connection with the grid. Areas under construction also need power supply before they are connected. PV systems are an attractive option for these cases. Furthermore, PV systems can be backed up by conventional generators to provide uninterrupted supply.

Remote area in Africa powered by PV.

Portable PV systems powering area struck by natural disaster.

Disaster Relief

Natural calamities often bring about an electricity crisis. As the disasters such as hurricanes, floods, tornadoes, and earthquakes destroy electricity generation and distribution systems. In situations like these, where power will be out for an extended period, portable PV systems can provide temporary solutions for light, communication, food and water systems. Emergency health clinics opt for PV based electricity over conventional systems in lieu to problems of fuel transport and pollution.

Scientific Experiments

In various cases, scientific experiments are set up in areas far from power supply. PV systems can be effectively used to carry out scientific activities in remote areas. Systems monitoring seismic activities, highway conditions, meteorological information and other research activities can be powered by PV systems.

Remote scientific experiment powered by PV.

Signal Systems

Navigational systems, such as light houses, highway and aircraft warning signals can be far from the electric grid. PV systems can be a reliable power source for these important applications. Even portable traffic lights can be powered by PV systems.

Light house being powered by PV.

Portable traffic lights powered by PV.

Water Pumping

PV is a perfect candidate for agricultural and livestock purposes due to the need for water during the periods with bright sunshine. These pumping systems can supply water directly to fields, or can store water for the time of need. These systems can even be used to provide water to remote areas and villages.

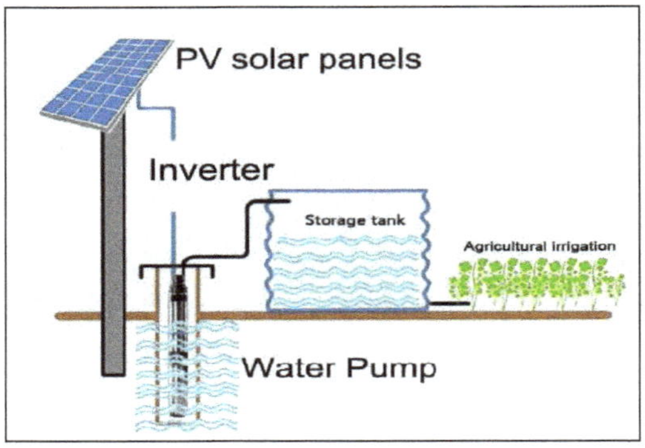

Solar water pumps can be very cost effective for remote agricultural activities.

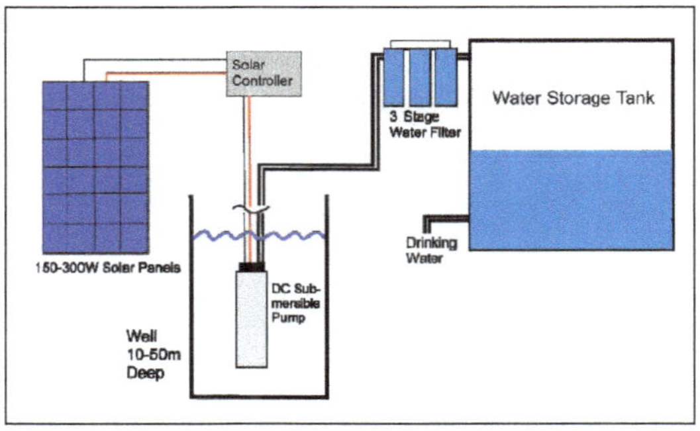

Solar water pumps can supply water to areas with no connection to grid.

Charging Vehicle Batteries

Vehicles running on electric power can be charged at PV powered stations. Such vehicles can also maintain their critical battery states using PV powered sources. Boats and other leisure vehicles can be charged directly using PV systems.

Electric vehicles at PV charging station.

Boat with a PV charging system.

Solar Power Cathodic Protection

Pipelines, well heads and other metallic structures are prone to corrosion due to exposure to water. Corrosion occurs due to the electrolytic activity of metals as they lose ions in contact with water. This electrolytic process leading to corrosion can be reduced by applying an external voltage. This external voltage will prevent the ion loss from the metal. To that end, only a small DC voltage will be enough. PV are a suitable candidate for this purpose as they produce low voltage DC power that can be used directly.

Cathodic protection using PV.

Refrigeration

PV system can be exceptionally suitable for storage and transport of medicines and vaccines that require refrigeration.

Solar powered refrigerator.

Consumer Products

PV technology is being used for variety of commercially available consumer based products. Small DC appliances such as toys, watches, calculators, radios, televisions, flashlights, fans etc. can operate with PV based energy systems.

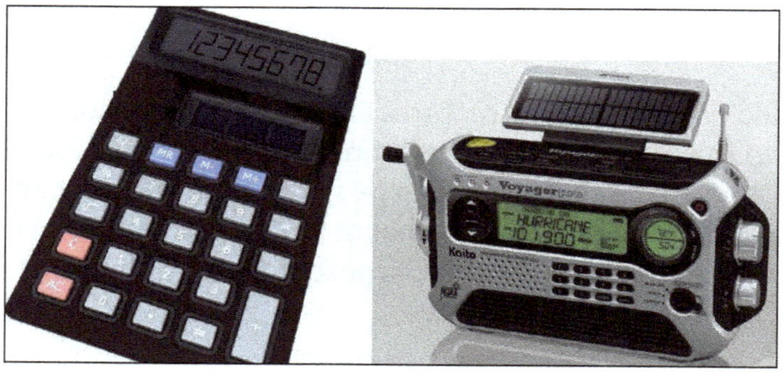

Solar powered calculator and radio.

Public Utilities

Various public utility systems such as teller machines and telephone booths can also be powered by PV systems.

ATM and telephone booth powered by PV.

The huge amount of research that has been carried over the past decade has led to an increased interest in implementing PV systems to satisfy energy needs. PV systems can be mounted directly to the building structure instead of putting on separate support structures, thus reducing space requirements. PV systems are available on flexible modules and can be mounted on almost every surface to harness solar energy. Finally, we should mention that, with the advance of technology, transparent and wearable PV systems are not far.

CONCENTRATED SOLAR POWER

The term Concentrating Solar Power (CSP) covers a range of technologies that utilize optical devices, such as mirrors and lenses, to concentrate the beam solar radiation and to provide for higher efficiency of its conversion into other forms of energy. Typically in many sources, CSP systems are associated with the solar thermal power; although, in a general sense, CSP can work with both thermal solar power and photovoltaic applications. Conversion of the concentrated sunlight can follow three routes: (1) conversion to heat, (2) conversion to electricity, and (3) conversion to fuels. The large utility scale plants primarily use the concentration of thermal energy, which is used to operate a steam turbine generator and produce electric power on site. In this case, the solar heat is used as any other source of heat (such as coal combustion, etc.) to generate steam.

CSP systems can only benefit from the direct beam radiation and therefore are best suited to the regions with a high percentage of clear sky days. The locations that have significant cloudiness, smog, or dust are not favorable.

By concentrator configuration, the commercial CSP systems are represented by:

- Parabolic trough systems.
- Central receiver towers.
- Parabolic dish systems.
- Linear Fresnel concentrators.

The CSP technology is one of the competitive options in energy industry for combustion-free electricity generation. Because the fuel cost is zero, the cost of the CSP technology is mainly associated with the significant initial capital investment. However, cost reduction trend for CSP is confidently predicted in the near future. The main avenues for cost efficiency are linked to:

- Technical progress in CSP technology, research and development efforts, and lesson learned from operating plants.
- Scaling up to larger plant size, which allows for the use of more cost effective turbines for power conversion.

The CSP technology has been commercialized and has experienced rapid growth since 2005, in part stimulated by the international concern for increased fossil fuel combustion and climate change. Worldwide, the CSP expansion was led by Spain, which has most CSP plants currently installed.

Linear Concentrator System for Concentrating Solar Power

Linear concentrating solar power (CSP) collectors capture the sun's energy with large mirrors that reflect and focus the sunlight onto a linear receiver tube. The receiver contains a fluid that is heated by the sunlight and then used to heat a traditional power cycle that spins a turbine that drives a generator to produce electricity. Alternatively, steam can be generated directly in the solar field, which eliminates the need for costly heat exchangers.

Linear concentrating collector fields consist of a large number of collectors in parallel rows that are typically aligned in a north-south orientation to maximize annual and summer energy collection. With a single-axis sun-tracking system, this configuration enables the mirrors to track the sun from east to west during the day, which ensures that the sun reflects continuously onto the receiver tubes.

Linear systems may incorporate thermal storage. In these systems, the collector field is oversized to heat a storage system during the day so the additional steam it generates can be used to produce electricity in the evening or during cloudy weather. These plants can also be designed as hybrids, meaning that they use fossil fuel to supplement the solar output during periods of low solar radiation. In such a design, a natural gas-fired heater or gas-steam boiler/reheater is used. In the future, linear systems may be integrated with existing or new combined-cycle natural-gas- and coal-fired plants.

Parabolic Trough Systems

The most common CSP system in the United States is a linear concentrator that uses parabolic trough collectors. In such a system, the receiver tube is positioned along the focal line of each parabola-shaped reflector. The tube is fixed to the mirror structure and the heat transfer fluid flows through and out of the field of solar mirrors to where it is used to create steam (or, in the case of a water/steam receiver, it is sent directly to the turbine).

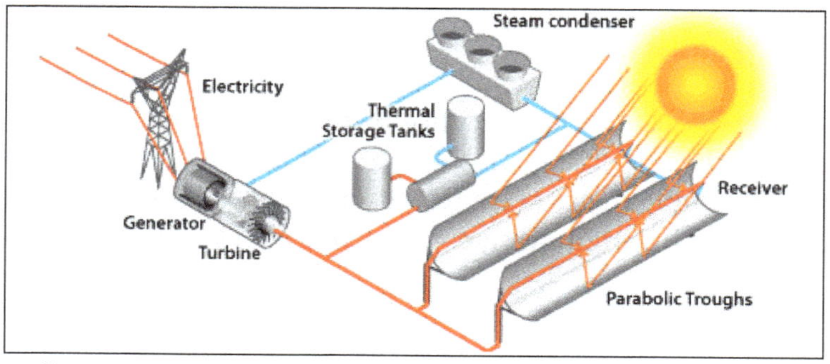

A linear concentrator power plant using parabolic trough collectors.

Linear Fresnel Reflector Systems

A second linear concentrator technology is the linear Fresnel reflector system. Flat or slightly curved mirrors mounted on trackers on the ground are configured to reflect sunlight onto a receiver tube fixed in space above the mirrors. A small parabolic mirror is sometimes added atop the receiver to further focus the sunlight.

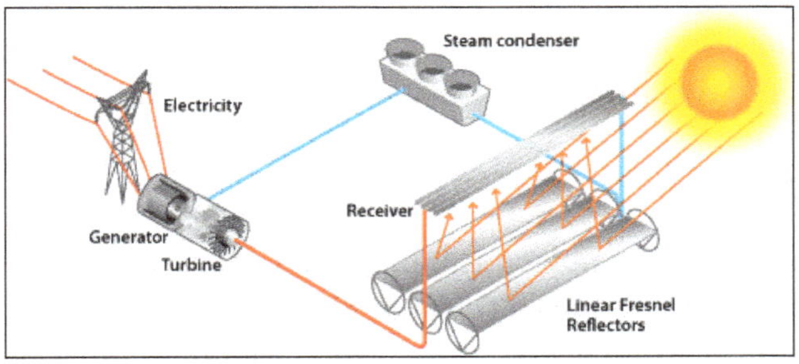

A linear Fresnel reflector power plant.

Dish/Engine System

Dish/engine systems use a parabolic dish of mirrors to direct and concentrate sunlight onto a central engine that produces electricity. The dish/engine system is a concentrating solar power (CSP) technology that produces smaller amounts of electricity than other CSP technologies—typically in the range of 3 to 25 kilowatts—but is beneficial for modular use. The two major parts of the system are the solar concentrator and the power conversion unit.

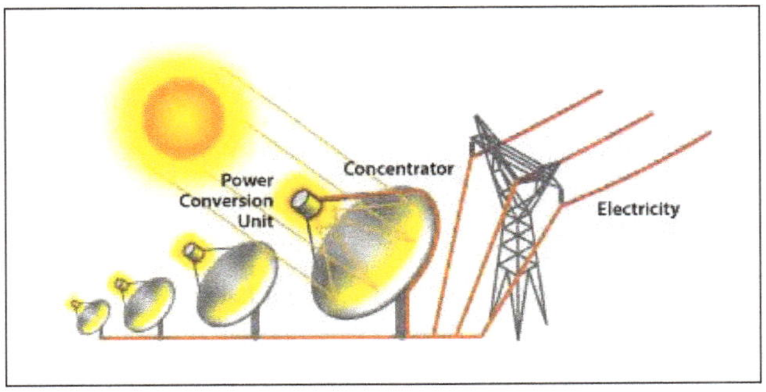

A dish/engine power plant.

Solar Concentrator

The solar concentrator, or dish, gathers the solar energy coming directly from the sun. The resulting beam of concentrated sunlight is reflected onto a thermal receiver that collects the solar heat. The dish is mounted on a structure that tracks the sun continuously throughout the day to reflect the highest percentage of sunlight possible onto the thermal receiver.

Power Conversion Unit

The power conversion unit includes the thermal receiver and the engine/generator. The thermal receiver is the interface between the dish and the engine/generator. It absorbs the concentrated beams of solar energy, converts the energy to heat, and transfers the heat to the engine/generator. A thermal receiver can be a bank of tubes with a cooling fluid—usually hydrogen or helium—that typically is the heat-transfer medium and also the working fluid for an engine. Alternate thermal receivers are heat pipes, where the boiling and condensing of an intermediate fluid transfers the heat to the engine.

The engine/generator system is the subsystem that takes the heat from the thermal receiver and uses it to produce thermal to electric energy conversion. The most common type of heat engine used in dish/engine systems is the Stirling engine. A Stirling engine uses the heated fluid to move pistons and create mechanical power. The mechanical work, in the form of the rotation of the engine's crankshaft, drives a generator and produces electrical power.

Power Tower System

In power tower concentrating solar power systems, a large number of flat, sun-tracking mirrors, known as heliostats, focus sunlight onto a receiver at the top of a tall tower. A heat-transfer fluid heated in the receiver is used to heat a working fluid, which, in turn, is used in a conventional turbine generator to produce electricity. Some power towers use water/steam as the heat-transfer fluid. Other advanced designs are experimenting with high temperature molten salts or sand-like particles to maximize the power cycle temperature.

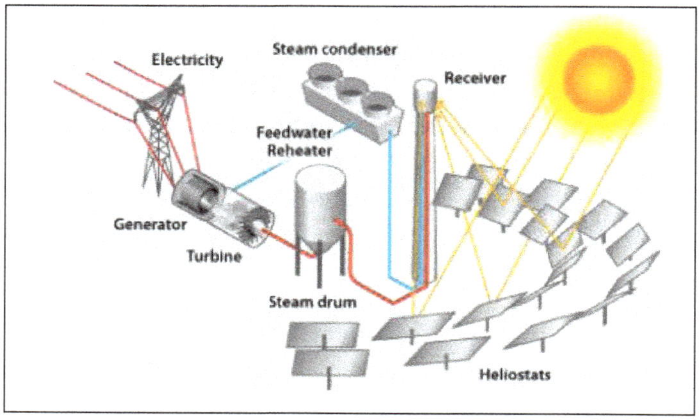

A power tower CSP plant.

Two large-scale power tower projects are currently operating in the U.S. The Ivanpah Solar Electric Generating System is the largest concentrated solar thermal plant in the country. Located in California's Mojave Desert, the plant is capable of producing 392 megawatts of electricity using 173,500 heliostats, each with two mirrors that focus sunlight onto three solar power towers. Crescent Dunes Solar Energy Project is a 110 megawatt solar thermal power plant located in southwestern Nevada. The project includes 10,347 heliostats and is the largest power tower system with direct thermal energy storage, which means the heat transfer fluid is also the storage material.

Aside from the U.S., Spain has several power tower systems. Planta Solar 10 and Planta Solar 20 are water/steam systems with capacities of 11 and 20 megawatts, respectively. Gemasolar, previously known as Solar Tres, produces nearly 20 megawatts of electricity and utilizes molten-salt thermal storage.

Thermal Storage System

One challenge facing the widespread use of solar energy is reduced or curtailed energy production when the sun sets or is blocked by clouds. Thermal energy storage provides a workable solution to this challenge.

In a concentrating solar power (CSP) system, the sun's rays are reflected onto a receiver, which creates heat that is used to generate electricity that can be used immediately or stored for later use. This enables CSP systems to be flexible, or dispatchable, options for providing clean, renewable energy.

Several sensible thermal energy storage technologies have been tested and implemented since 1985. These include the two-tank direct system, two-tank indirect system, and single-tank thermocline system.

Two-tank Direct System

Solar thermal energy in this system is stored in the same fluid used to collect it. The fluid is stored in two tanks—one at high temperature and the other at low temperature. Fluid from the low-temperature tank flows through the solar collector or receiver, where solar energy heats it to a high temperature, and it then flows to the high-temperature tank for storage. Fluid from the high-temperature tank flows through a heat exchanger, where it generates steam for electricity production. The fluid exits the heat exchanger at a low temperature and returns to the low-temperature tank.

The two-tank direct molten-salt thermal energy storage system at the Solar Two power plant.

Two-tank direct storage was used in early parabolic trough power plants (such as Solar Electric Generating Station I) and at the Solar Two power tower in California. The trough plants used mineral oil as the heat-transfer and storage fluid; Solar Two used molten salt.

Two-tank Indirect System

Two-tank indirect systems function in the same way as two-tank direct systems, except different fluids are used as the heat-transfer and storage fluids. This system is used in plants in which the heat-transfer fluid is too expensive or not suited for use as the storage fluid.

The two-tank indirect thermal energy storage system for AndaSol-1 and -2.

The storage fluid from the low-temperature tank flows through an extra heat exchanger, where it is heated by the high-temperature heat-transfer fluid. The high-temperature storage fluid then flows back to the high-temperature storage tank. The fluid exits this heat exchanger at a low temperature and returns to the solar collector or receiver, where it is heated back to a high temperature. Storage fluid from the high-temperature tank is used to generate steam in the same manner as the two-tank direct system. The indirect system requires an extra heat exchanger, which adds cost to the system.

This system will be used in many of the parabolic power plants in Spain and has also been proposed for several U.S. parabolic plants. The plants will use organic oil as the heat-transfer fluid and molten salt as the storage fluid.

Single-tank Thermocline System

Single-tank thermocline systems store thermal energy in a solid medium—most commonly, silica sand—located in a single tank. At any time during operation, a portion of the medium is at high temperature, and a portion is at low temperature. The hot- and cold-temperature regions are separated by a temperature gradient or thermocline. High-temperature heat-transfer fluid flows into the top of the thermocline and exits the bottom at low temperature. This process moves the thermocline downward and adds thermal energy to the system for storage. Reversing the flow moves the thermocline upward and removes thermal energy from the system to generate steam and electricity. Buoyancy effects create thermal stratification of the fluid within the tank, which helps to stabilize and maintain the thermocline.

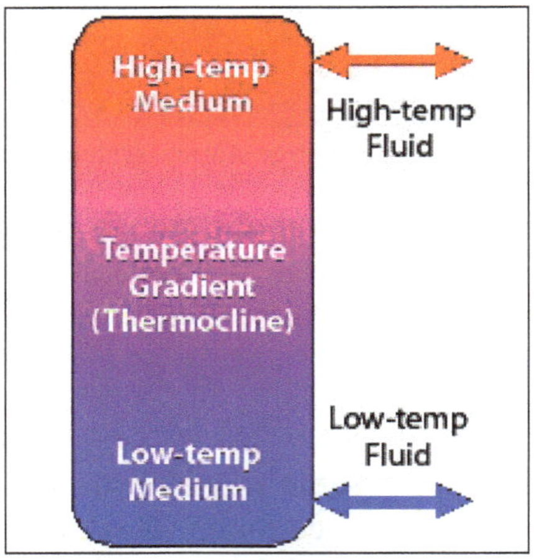

A single-tank thermocline thermal energy storage system.

Using a solid storage medium and only needing one tank reduces the cost of this system relative to two-tank systems. This system was demonstrated at the Solar One power tower, where steam was used as the heat-transfer fluid and mineral oil was used as the storage fluid.

WORKING OF CONCENTRATED SOLAR POWER

Concentrating solar power plants produce electric power by converting the sun's energy into high-temperature heat using various mirror configurations. The heat is then channelled through a conventional generator. The plants consist of two parts: one that collects solar energy and converts it to heat, and another that converts heat energy to electricity.

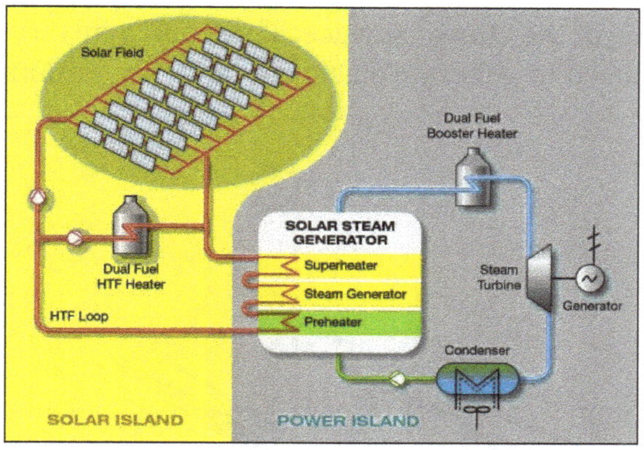

Concentrating solar power systems can be sized for village power (10 kilowatts) or grid-connected applications (up to 100 megawatts). Some systems use thermal storage during cloudy periods or at night. Others can be combined with natural gas and the resulting hybrid power plants provide high-value, dispatchable power. These attributes, along with world record solar-to-electric conversion efficiencies, make concentrating solar power an attractive renewable energy option in the Southwest and other sunbelt regions worldwide. There are four CSP technologies being promoted internationally. For each of these, there exists various design variations or different configurations. The amount of power generated by a concentrating solar power plant depends on the amount of direct sunlight. Like concentrating photovoltaic concentrators, these technologies use only direct-beam sunlight, rather than diffuse solar radiation.

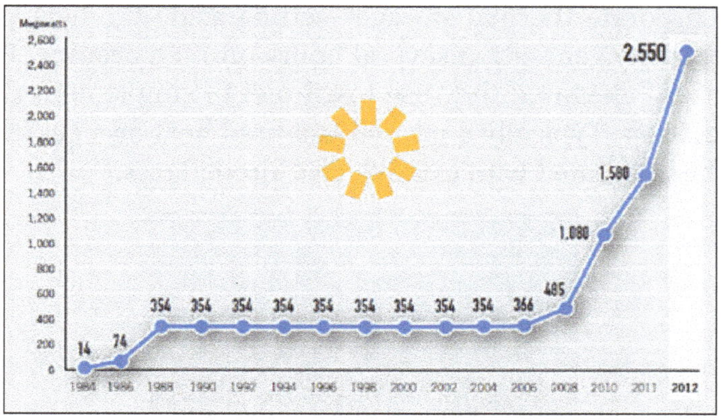

The 10 MW Solar One power tower was developed in Southern California in 1981, but the parabolic-trough technology of the nearby Solar Energy Generating Systems (SEGS), begun in 1984, was more workable. The 354 MW SEGS is still the largest solar power plant in the world, and will remain so until the 390 MW Ivanpah power tower project comes online.

Concentrated Solar Power (MW)												
Year	1984	1985	1989	1990	...	2006	2007	2008	2009	2010	2011	2012
Installed	14	60	200	80	0	1	74	55	178.50	306.50	628.5	802.5
Cumulative	14	74	274	354	354	355	429	484	662.5	969	1597.5	2553

Parabolic Trough Systems

The sun's energy is concentrated by parabolically curved, trough-shaped reflectors onto a receiver pipe running along the inside of the curved surface. This energy heats oil flowing through the pipe, and the heat energy is then used to generate electricity in a conventional steam generator.

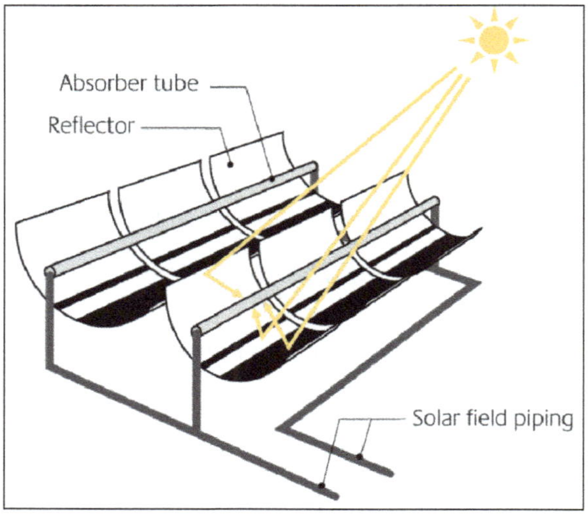

A collector field comprises many troughs in parallel rows aligned on a north-south axis. This configuration enables the single-axis troughs to track the sun from east to west during the day to ensure that the sun is continuously focused on the receiver pipes. Individual trough systems currently can generate about 80 megawatts of electricity.

Trough designs can incorporate thermal storage—setting aside the heat transfer fluid in its hot phase—allowing for electricity generation several hours into the evening. Currently, all parabolic trough plants are "hybrids," meaning they use fossil fuel to supplement the solar output during periods of low solar radiation. Typically a natural gas-fired heat or a gas steam boiler/reheater is used; troughs also can be integrated with existing coal-fired plants.

Another option under investigation is the approximation of the parabolic troughs by segmented mirrors according to the principle of Fresnel.

Power Tower Systems

A power tower converts sunshine into clean electricity for the world's electricity grids. The technology utilizes many large, sun-tracking mirrors (heliostats) to focus sunlight on a receiver at the top of a tower. A heat transfer fluid heated in the receiver is used to generate steam, which, in turn, is used in a conventional turbine-generator to produce electricity.

Early power towers (such as the Solar One plant) utilized steam as the heat transfer fluid; current US designs (including Solar Two, pictured) utilize molten nitrate salt because of its superior heat transfer and energy storage capabilities.

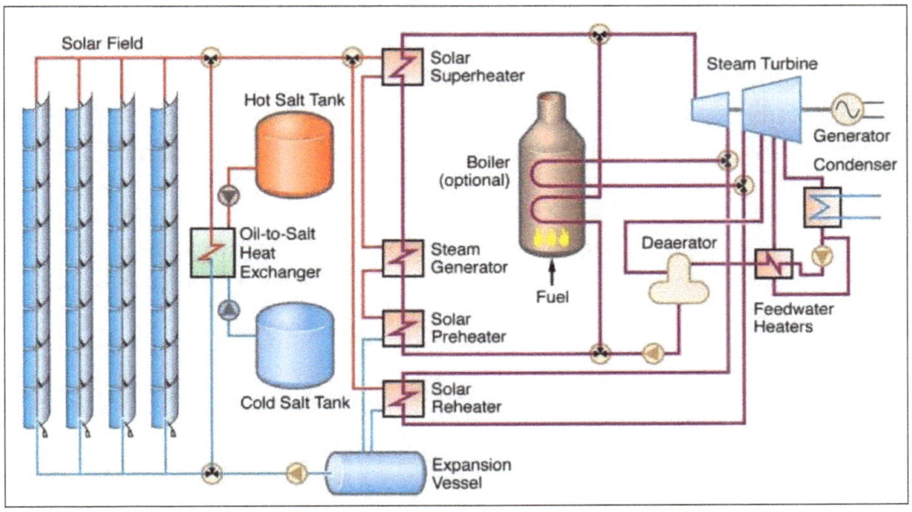

Concentrated Solar Power - Power Tower Process Digram.

Current European designs use air as heat transfer medium because of its high temperature and its good handability.

Individual commercial plants will be sized to produce anywhere from 50 to 200 MW of electricity.

Solar Dish Systems

Parabolic dish systems consist of a parabolic-shaped point focus concentrator in the form of a dish that reflects solar radiation onto a receiver mounted at the focal point. These concentrators are mounted on a structure with a two-axis tracking system to follow the sun.

The collected heat is typically utilized directly by a heat engine mounted on the receiver moving with the dish structure. Stirling and Brayton cycle engines are currently favored for power conversion. Projects of modular systems have been realized with total capacities up to 5 MWe. The modules have maximum sizes of 50 kWe and have achieved peak efficiencies up to 30% net.

Pros and Cons of CSP

Concentrated solar power or CSP is an alternative and renewable energy technology centered on indirect conversion of sunlight into electricity. Unlike solar power through photovoltaic solar panels that directly convert radiant energy from the sun into electricity, CSP uses an array of mirrors placed in a large area of land to direct and concentrate sunlight into a single point known as the receiver. The receiver collects the aggregate heat. This heat is used to drive a steam turbine connected to an electrical power generator or incite a thermonuclear reaction.

Nonetheless, similar to photovoltaic solar power and other alternative energy technologies such as wind power and hydropower, concentrated solar power has an advantage of being a renewable, sustainable or self-sufficient, and clean source of energy. Note it has other advantages, as well as disadvantages.

Advantages of Concentrated Solar Power

Uncomplicated Implementations and Operations

One of the remarkable benefits or advantages of concentrated solar power is that its corresponding power plant closely resembles most power plants based on steam turbines. Plants running on fossil fuels can technically be used for CSP systems. The operating cost of a CSP plant is comparatively lower than hydrocarbon-based and nuclear-based plants because of simpler operations and maintenance requirements. Depending on the location and the size of the entire plant, a CSP system is scalable up to more than 100 MW level.

Supplements other Sources of Energy

Building a CSP plant can complement other sources of energy, thus promoting a more secure energy grid. Although this renewable energy source does not generate electricity enough to power large communities, adding it to the energy mix can address existing and future electricity requirements. CSP can also be used in effective and efficient oil recovery. Through so-called solar thermal enhanced oil recovery, the steam it produces is used to make heavy oil less viscous and easier to pump, thus extending the lives of fields with heavy oil and making them more economical.

Relatively Uninterrupted Source of Electricity

Round-the-clock generation of electricity is another remarkable advantage of concentrated solar power technology, especially when compared to photovoltaic solar panel and wind power technologies. Take note that photovoltaic solar panels and wind power are intermittent in nature. On the other hand, certain CSP plants can store energy in the form of molten salts. Hence, the electricity they generate is more sustainable, predictable, and reliable.

Converts Solar Energy into a Transportable Form Energy

Concentrated solar power has other uses beyond electricity generation. Several institutions are investigating and improving the application of CSP in thermal reaction technologies for solar fuel production. For example, a couple of studies have demonstrated the use of CSP to break apart water molecules to harness hydrogen without carbon emissions. The study of Elena Diaz et al. also demonstrated the use of CSP to decompose methane into hydrogen and carbon fuel cells. Promoting these processes further can make solar a transportable form of energy.

Disadvantages of Concentrated Solar Power

Dependent on Locations and Large Tracks of Lands

Similar to photovoltaic solar power and wind power, a fundamental limitation or disadvantage of concentrated solar power is that it requires using extensive land area that otherwise, could be used for commercial and residential development or agriculture. In addition, the study of M. Enjavi-Arsanji, K. Hibordi, and M. Yaghoubi noted that a CSP plant is economical only for locations with direct normal irradiation beyond 1800 kWh/m^2/year such as sprawling open spaces and desert areas. This technology is not economical in populated areas and regions that receive less sunlight in a year.

Notable and Possible Negative Ecological Impacts

Another drawback of concentrated solar power is that it uses a lot of water either to drive steam turbines for electricity generation or to cool down thermochemical reactors. Although using sea water can be a conventional solution, this would imply building a CSP plant near the coastline that in turn, may not be suitable depending on the amount of solar radiation the specific area receives. There is also another issue concerning the impacts of CSPs on wildlife. Because it directs and redirects sunlight, it naturally attracts certain species such as insects that are preys of other species such as birds. The level of heat around the area can kill predators.

Costs Implications of Thermal Storage Materials

The study of Jun Wang et al. that several factors influence the costs of building and maintaining a CSP plant. The use of steam or thermal oil is not an attractive option for high-temperature thermal energy storage material because of its immediate cost and other costs associated with its temperature limitations and flammability. Molten salt is an alternative storage material but aside from drawback due to limited operating temperature range because it solidifies at low temperature and decomposes at high temperature, it promotes corrosion and deterioration, which in turn, requires the use of more expensive storage tank materials.

Possible Lack of Attention from Researchers and Investors

Different types of renewable and alternative sources of energy are competing against one another in terms of research attention and investment interest. Concentrated solar power is competing with photovoltaic solar power and wind power. Breakthroughs in photovoltaic technologies have increased the cost and energy efficiency of solar panels. Take note that CSP is also competing against more efficient sources of energy such as fission-based nuclear power. After all, CSP is not as energy dense as nuclear power and hydrocarbons. Further developments in the future might render CSP obsolete.

CONCENTRATOR PHOTOVOLTAICS

This Amonix system in Las Vegas, USA consists of thousands of small Fresnel lenses, each focusing sunlight to ~500X higher intensity onto a tiny, high-efficiency multi-junction solar cell. A Tesla Roadster is parked beneath for scale.

Concentrator photovoltaics (CPV) (also known as concentration photovoltaics) is a photovoltaic technology that generates electricity from sunlight. Unlike conventional photovoltaic systems, it uses lenses or curved mirrors to focus sunlight onto small, highly efficient, multi-junction (MJ) solar cells. In addition, CPV systems often use solar trackers and sometimes a cooling system to further increase their efficiency. Ongoing research and development is rapidly improving their competitiveness in the utility-scale segment and in areas of high insolation.

Concentrator photovoltaics (CPV) modules on dual axis solar trackers.

Systems using high-concentration photovoltaics (HCPV) especially have the potential to become competitive in the near future. They possess the highest efficiency of all existing PV technologies, and a smaller photovoltaic array also reduces the balance of system costs. Currently, CPV is not used in the PV rooftop segment and is far less common than conventional PV systems. For regions with a high annual direct normal irradiance of 2000 kilowatt-hour (kWh) per square meter or more, the levelized cost of electricity is in the range of \$0.08–\$0.15 per kWh and installation cost for a 10-megawatt CPV power plant was identified to lie between €1.40–€2.20 (~\$1.50-\$2.30) per watt-peak (W_p).

In 2016, cumulative CPV installations reached 350 megawatts (MW), less than 0.2% of the global installed capacity of 230,000 MW. Commercial HCPV systems reached instantaneous ("spot") efficiencies of up to 42% under standard test conditions (with concentration levels above 400) and the International Energy Agency sees potential to increase the efficiency of this technology to 50% by the mid-2020s. As of December 2014, the best lab cell efficiency for concentrator MJ-cells reached 46% (four or more junctions). Under outdoor, operating conditions, CPV module efficiencies have exceeded 33% ("one third of a sun"). System-level AC efficiencies are in the range of 25-28%. CPV installations are located in China, the United States, South Africa, Italy and Spain.

HCPV directly competes with concentrated solar power (CSP) as both technologies are suited best for areas with high direct normal irradiance, which are also known as the Sun Belt region in the United States and the Golden Banana in Southern Europe. CPV and CSP are often confused with one another, despite being intrinsically different technologies from the start: CPV uses the photovoltaic effect to directly generate electricity from sunlight, while CSP – often called concentrated solar thermal – uses the heat from the sun's radiation in order to make steam to drive a turbine, that then produces electricity using a generator. As of 2012, CSP is still more common than CPV.

Challenges

Modern CPV systems operate most efficiently in highly concentrated sunlight (i.e. concentration levels equivalent to hundreds of suns), as long as the solar cell is kept cool through the use of heat sinks. Diffuse light, which occurs in cloudy and overcast conditions, cannot be highly concentrated using conventional optical components only (i.e. macroscopic lenses and mirrors). Filtered light, which occurs in hazy or polluted conditions, has spectral variations which produce mismatches between the electrical currents generated within the series-connected junctions of spectrally "tuned" multi-junction (MJ) photovoltaic cells. These CPV features lead to rapid decreases in power output when atmospheric conditions are less than ideal.

To produce equal or greater energy per rated watt than conventional PV systems, CPV systems must be located in areas that receive plentiful direct sunlight. This is typically specified as average DNI greater than 5.5-6 kWh/m²/day or 2000kWh/m²/yr. Otherwise, evaluations of annualized DNI vs. GNI/GHI irradiance data have concluded that conventional PV should still perform better over time than presently available CPV technology in most regions of the world.

Ongoing Research and Development

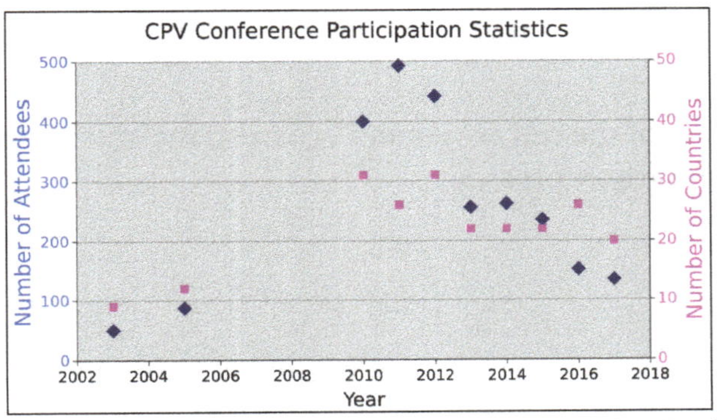

International CPV-x Conference - Historical Participation Statistics.

CPV research and development has been pursued in over 20 countries for more than a decade. The annual CPV-x conference series has served as a primary networking and exchange forum between university, government lab, and industry participants. Government agencies have also continued to encourage a number of specific technology thrusts.

ARPA-E announced a first round of R&D funding in late 2015 for the MOSAIC Program (Microscale Optimized Solar-cell Arrays with Integrated Concentration) to further combat the location and expense challenges of existing CPV technology. As stated in the program description: "MOSAIC projects are grouped into three categories: complete systems that cost effectively integrate micro-CPV for regions such as sunny areas of the U.S. southwest that have high Direct Normal Incident (DNI) solar radiation; complete systems that apply to regions, such as areas of the U.S. Northeast and Midwest, that have low DNI solar radiation or high diffuse solar radiation; and concepts that seek partial solutions to technology challenges."

In Europe the CPVMATCH Program (Concentrating PhotoVoltaic Modules using Advanced Technologies and Cells for Highest efficiencies) aims "to bring practical performance of HCPV modules closer to theoretical limits". Efficiency goals achievable by 2019 are identified as 48% for cells and 40% for modules at >800x concentration. A 41.4% module efficiency was announced at the end of 2018.

The Australian Renewable Energy Agency (ARENA) extended its support in 2017 for further commercialization of the HCPV technology developed by Raygen. Their 250 kW dense array receivers are the most powerful CPV receivers thus far created, with demonstrated PV efficiency of 40.4% and include usable heat co-generation.

Efficiency

According to theory, semiconductor properties allow solar cells to operate more efficiently in concentrated light than they do under a nominal level of solar irradiance. This is because, along with a proportional increase in the generated current, there also occurs a logarithmic enhancement in operating voltage, in response to the higher illumination.

To be explicit, consider the power (P) generated by a solar cell under "one-sun" illumination at the earth's surface, which corresponds to a peak solar irradiance Q=1000 Watts/m^2. The cell power can be expressed as a function of the open-circuit voltage (V_{oc}), the short-circuit current (I_{sc}), and the fill factor (FF) of the cell's characteristic current-voltage (I-V) curve:

$$P = I_{sc} \times V_{oc} \times FF.$$

Upon increased illumination of the cell at "χ-suns", corresponding to concentration (χ) and irradiance (χQ), there can be similarly expressed:

$$P_{\chi} = I_{\chi sc} \times V_{\chi oc} \times FF_{\chi}$$

where, as shown by reference:

$$I_{\chi sc} = \chi \times I_{sc} \text{ and } V_{\chi oc} = V_{oc} + \frac{kT}{q} \ln(\chi).$$

Note that the unitless fill factor for a "high quality" solar cell typically ranges 0.75-0.9 and can, in practice, depend primarily on the equivalent shunt and series resistances for the particular cell construction. For concentrator applications, FF and FF_χ should then have similar values that are both near unity, corresponding to high shunt resistance and very low series resistance (<1 milliohm).

The efficiencies of a cell of area (A) under one-sun and χ-suns are defined as:

$$\eta = \frac{P}{QA} \text{ and } \eta_\chi = \frac{P_\chi}{\chi QA}.$$

The efficiency under concentration is then given in terms of χ and the cell characteristics as:

$$\eta_\chi = \eta \times \frac{P_\chi}{\chi P} = \eta \times \left(1 + \frac{kT}{q}\frac{\ln(\chi)}{V_{oc}}\right) \times \left(\frac{FF_\chi}{FF}\right)$$

where, the term kT/q is the voltage (called the thermal voltage) of a thermalized population of electrons - such as that flowing through a solar cell's p-n junction - and has a value of about 25.85 mV at room temperature (300 K).

The efficiency enhancement of η_χ relative to η is listed in the following table for a set of typical open-circuit voltages that roughly represent different cell technologies. The table shows that the enhancement can be as much as 20-30% at χ = 1000 concentration. The calculation assumes FF_χ/FF=1; an assumption which is clarified in the following discussion.

Theoretical Cell Efficiency Increase Due to Sunlight Concentration			
Cell Technology	Multi-crystal Silicon	Mono-crystal Silicon	Triple-junction III-V on GaAs
Approximate Junction V_{oc}	550 mV	700 mV	850 mV
χ = 10	10.8%	8.5%	7.0%
χ = 100	21.6%	17.0%	14.0%
χ = 1000	32.5%	25.5%	21.0%

In practice, the higher current densities and temperatures which arise under sunlight concentration may be challenging to prevent from degrading the cell's I-V properties or, worse, causing permanent physical damage. Such effects can reduce the ratio FF_χ/FF by an even larger percentage below unity than the tabulated values shown above. To prevent irreversible damage, the rise in cell operating temperature under concentration must be controlled with the use of a suitable heat sink. Additionally, the cell design itself must incorporate features that reduce recombination and the contact, electrode, and busbar resistances to levels that accommodate the target concentration and resulting current density. These features include thin, low-defect semiconductor layers; thick, low-resistivity electrode & busbar materials; and small (typically <1 cm^2) cell sizes.

Including such features, the best thin film multi-junction photovoltaic cells developed for terrestrial CPV applications achieve reliable operation at concentrations as high as 500-1000 suns (i.e. irradiances of 50-100 Watts/cm^2). As of year 2014, their efficiencies are upwards of 44% (three junctions), with the potential to approach 50% (four or more junctions) in the coming years . The theoretical limiting efficiency under concentration approaches 65% for 5 junctions, which is a likely practical maximum.

Optical Design

All CPV systems have a solar cell and a concentrating optic. Optical sunlight concentrators for CPV introduce a very specific design problem, with features that make them different from most other optical designs. They have to be efficient, suitable for mass production, capable of high concentration, insensitive to manufacturing and mounting inaccuracies, and capable of providing uniform illumination of the cell. All these reasons make nonimaging optics the most suitable for CPV.

For very low concentrations, the wide acceptance angles of nonimaging optics avoid the need for active solar tracking. For medium and high concentrations, a wide acceptance angle can be seen as a measure of how tolerant the optic is to imperfections in the whole system. It is vital to start with a wide acceptance angle since it must be able to accommodate tracking errors, movements of the system due to wind, imperfectly manufactured optics, imperfectly assembled components, finite stiffness of the supporting structure or its deformation due to aging, among other factors. All of these reduce the initial acceptance angle and, after they are all factored in, the system must still be able to capture the finite angular aperture of sunlight.

Types

CPV systems are categorized according to the amount of their solar concentration, measured in "suns" (the square of the magnification).

Low Concentration PV (LCPV)

Low concentration PV are systems with a solar concentration of 2–100 suns. For economic reasons, conventional or modified silicon solar cells are typically used. The heat flux is typically low enough that the cells do not need to be actively cooled. For standard solar modules, there is also modeling and experimental evidence that no tracking or cooling modifications are needed if the concentration level is low.

An example of a Low Concentration PV Cell's surface, showing the glass lensing.

Low-concentration systems often have a simple booster reflector, which can increase solar electric output by over 30% from that of non-concentrator PV systems. Experimental results from such LCPV systems in Canada resulted in energy gains over 40% for prismatic glass and 45% for traditional crystalline silicon PV modules.

Medium Concentration PV

From concentrations of 100 to 300 suns, the CPV systems require two-axis solar tracking and cooling (whether passive or active), which makes them more complex.

A 10×10 mm HCPV solar cell.

High Concentration PV (HCPV)

High concentration photovoltaics (HCPV) systems employ concentrating optics consisting of dish reflectors or fresnel lenses that concentrate sunlight to intensities of 1,000 suns or more. The solar cells require high-capacity heat sinks to prevent thermal destruction and to manage temperature related electrical performance and life expectancy losses. To further exacerbate the concentrated cooling design, the heat sink must be passive, otherwise the power required for active cooling will reduce the overall conversion efficiency and economy. Multi-junction solar cells are currently favored over single junction cells, as they are more efficient and have a lower temperature coefficient (less loss in efficiency with an increase in temperature). The efficiency of both cell types rises with increased concentration; multi-junction efficiency rises faster. Multi-junction solar cells, originally designed for non-concentrating PV on space-based satellites, have been re-designed due to the high-current density encountered with CPV (typically 8 A/cm^2 at 500 suns). Though the cost of multi-junction solar cells is roughly 100 times that of conventional silicon cells of the same area, the small cell area employed makes the relative costs of cells in each system comparable and the system economics favor the multi-junction cells. Multi-junction cell efficiency has now reached 44% in production cells.

The 44% value given above is for a specific set of conditions known as "standard test conditions". These include a specific spectrum, an incident optical power of 850 W/m^2, and a cell temperature of 25 °C. In a concentrating system, the cell will typically operate under conditions of variable spectrum, lower optical power, and higher temperature. The optics needed to concentrate the light have limited efficiency themselves, in the range of 75–90%. Taking these factors into account, a solar module incorporating a 44% multi-junction cell might deliver a DC efficiency around 36%. Under similar conditions, a crystalline silicon module would deliver an efficiency of less than 18%.

When high concentration is needed (500–1000 times), as occurs in the case of high efficiency multi-junction solar cells, it is likely that it will be crucial for commercial success at the system level to achieve such concentration with a sufficient acceptance angle. This allows tolerance in mass production of all components, relaxes the module assembling and system installation, and decreasing the cost of structural elements. Since the main goal of CPV is to make solar energy inexpensive, there are only a few surfaces that can be used. Decreasing the number of elements and achieving high acceptance angle, can be relaxed optical and mechanical requirements, such as accuracy of the optical surfaces profiles, the module assembling, the installation, the supporting structure, etc. To this end, improvements in sunshape modelling at the system design stage may lead to higher system efficiencies.

Reliability

The higher capital costs, lesser standardization, and added engineering & operational complexities (in comparison to zero and low-concentration PV technologies) make long-life performance a critical demonstration goal for the first generations of CPV technologies. Performance certification standards (UL 3703,UL 8703, IEC 62108, IEC 62670, IEC 62789, and IEC 62817) include stress testing conditions that may be useful to uncover some predominantly infant and early life (<1–2 year) failure modes at the system, tracker, module, receiver, and other sub-component levels. However, such standardized tests – as typically performed on only a small sampling of units – are generally incapable to evaluate comprehensive long-term lifetimes (10 to 25 or more years) for each unique system design and application under its broader range of actual - and occasionally unanticipated - operating conditions. Reliability of these complex systems is therefore assessed in the field, and is improved through aggressive product development cycles which are guided by the results of accelerated component/system aging, performance monitoring diagnostics, and failure analysis. Significant growth in the deployment of CPV can be anticipated once the concerns are better addressed to build confidence in system bankability.

Tracker Durability and Maintenance

The tracker and module support structure for a modern HCPV system must each remain accurate within 0.1°-0.3° in order to keep the solar resource adequately centered within the acceptance angle of the receiver collection optics, and thus concentrated onto the PV cells. This is a challenging requirement for any mechanical system that is subjected to the stresses of varying movements and loads. Economical procedures for periodic realignment and maintenance of the tracker may thus be required to preserve system performance over its expected lifetime.

Receiver Temperature Control

The maximum multi-junction solar cell operating temperatures ($T_{max\ cell}$) of HCPV systems are limited to less than about 110 °C on account of their intrinsic reliability limitation. This contrasts to CSP and other CHP systems which may be designed to function at temperatures in excess of several hundred degrees. More specifically, the cells are fabricated from a layering of thin-film III-V semiconductor materials having intrinsic lifetimes during operation that rapidly decrease with an Arrhenius-type temperature dependence. The system receiver must therefore provide for highly efficient and uniform cell cooling through sufficiently robust active and/or passive methods. In addition to material and design limitations in receiver heat-transfer performance, other extrinsic factors - such as the frequent system thermal cycling - further reduce the practical $T_{max\ receiver}$ compatible with long system life to below about 80 °C.

Installations

Concentrator photovoltaics technology has established its presence in the solar industry over the last several years. The first HCPV power plant that exceeded 1 MW-level was commissioned in Spain in 2006. By the end of 2015, the number of CPV power plants (including both LCPV and HCPV) around the world accounted for a total installed capacity of 350 MW. Field data collected over six years is also starting to benchmark the prospects for long-term system reliability.

The emerging CPV segment has comprised ~0.1% of the fast-growing utility market for PV installations over the past decade. Unfortunately, following a rapid drop in traditional flat-panel PV prices, the near term outlook for CPV industry growth has faded as signaled by closure of the largest HCPV manufacturing facilities: including those of Suncore, Soitec, Amonix, and SolFocus. The higher cost and complexity of maintaining the precision HCPV dual-axis trackers has also been reported in some instances to be especially challenging. Nevertheless, the growth outlook for the PV industry as a whole continues to be strong, thus providing continued optimism that CPV technology will eventually demonstrate its place.

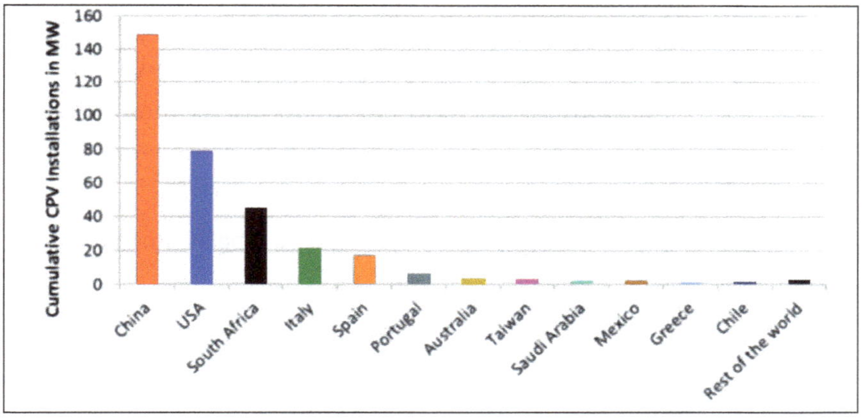

Cumulative CPV Installations in MW by country by November 2014.

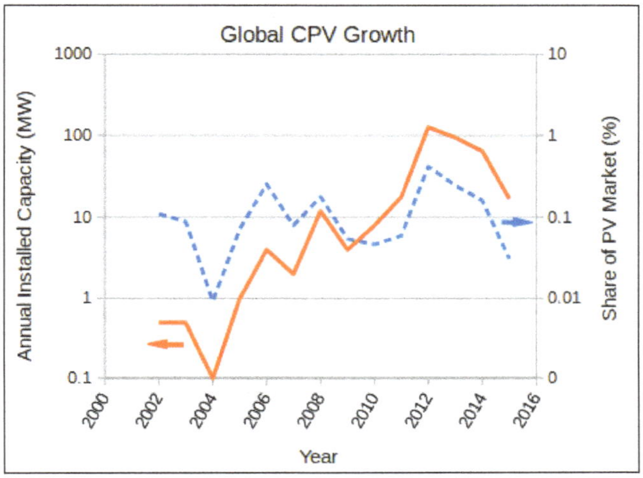

Yearly Installed CPV Capacity in MW from 2002 to 2015.

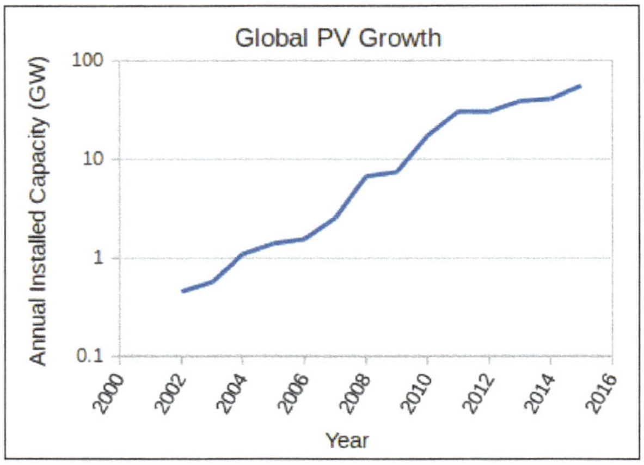

Yearly Installed PV Capacity in GW from 2002 to 2015.

List of Largest HCPV Systems

Similar to traditional PV, the peak DC rating of a system is specified as MW_p (or sometimes MW_{DC}) under concentrator standard test conditions (CSTC) of DNI=1000 W/m², AM1.5D, & T_{cell}=25 °C, as per the IEC 62670 standard convention. The AC production capacity is specified as MW_{AC} under IEC 62670 concentrator standard operating conditions (CSOC) of DNI=900 W/m², AM1.5D, $T_{ambient}$=20 °C, & Wind speed=2 m/s, and may include adjustments for inverter efficiency, higher/lower solar resource, and other facility-specific factors. The largest CPV power plant currently in operation is of 138 MW_p rating located in Golmud, China, hosted by Suncore Photovoltaics.

Power station	Rating (MW_p)	Capacity (MW_{AC})	Year Completed	Location	CPV Vendor
Golmud (1 and 2)	137.8	110	2012 - 2013	in Golmud/Qinghai province/China	Suncore
Touwsrivier CPV Project	44.2	36	2014	in Touwsrivier/Western Cape/ South Africa	Soitec
Alamosa Solar Project	35.3	30	2012	in Alamosa, Colorado/San Luis Valley/USA	Amonix
Hami (1, 2, and 3)	10.5	9.0	2013 - 2016	in Hami/Xinjiang province/China	Soitec-Focusic
Navarra CPV Plant	9.1	7.8	2010	in Villafranca/Navarra province/ Spain	Amonix-Guascor Foton

Table: List of HCPV systems in United States.

Power station	Rating (MW_p)	Capacity (MW_{AC})	Year Completed	Location	CPV Vendor	Owner/Operator
Alamosa Solar Project	35.3	30	2012	Alamosa, Colorado	Amonix	Cogentrix
Desert Green Solar Farm	7.80	6.3	2014	Borrego Spgs, California	Soitec	Invenergy
Hatch Solar Energy Center	5.88	5.0	2011	Hatch, New Mexico	Amonix	NextEra Energy
University of Arizona CPV Array	2.38	2.0	2011	Tucson, Arizona	Amonix	Arzon Solar

Newberry Springs CPV Power Plant	1.68	1.5	2013	Newberry Spgs, California	Soitec	STACE
Crafton Hills College Solar Farm	1.61	1.3	2012	Yucaipa, California	SolFocus	Crafton Hills College
Victor Valley College Solar Farm	1.26	1.0	2010	Victorville, California	SolFocus	Victor Valley College
Eubank Landfill Solar Array	1.21	1.0	2013	Albuquerque, New Mexico	Suncore	Emcore Solar
Questa Solar Facility	1.17	1.0	2010	Questa, New Mexico	Soitec	Chevron
Fort Irwin CPV Project	1.12	1.0	2015	Fort Irwin, California	Soitec	US DOD

Table: List of LCPV systems in United States.

Power station	Capacity (MW_{AC})	Year Completed	Location	Coordinates	CPV Vendor	Owner/Operator
Fort Churchill Solar Array	19.9	2015	Yerington, Nevada	39°07′41″N 119°08′24″W39.12806°N 119.14000°W	SunPower	Apple Inc./ NV Energy
Springerville Solar Farm	6.0	2013	Springerville, Arizona	34°17′40″N 109°16′17″W34.29444°N 109.27139°W	SunPower	Tucson Electric Power
ASU Polytechnic CPV Array	1.0	2012	Mesa, Arizona	33°17′37″N 111°40′38″W33.29361°N 111.67722°W	SunPower	SunPower

Concentrated Photovoltaics and Thermal

Concentrator photovoltaics and thermal (CPVT), also sometimes called combined heat and power solar (CHAPS) or hybrid thermal CPV, is a cogeneration or micro cogeneration technology used in the field of concentrator photovoltaics that produces usable heat and electricity within the same system. CPVT at high concentrations of over 100 suns (HCPVT) utilizes similar components as HCPV, including dual-axis tracking and multi-junction photovoltaic cells. A fluid actively cools the integrated thermal–photovoltaic receiver, and simultaneously transports the collected heat.

Typically, one or more receivers and a heat exchanger operate within a closed thermal loop. To maintain efficient overall operation and avoid damage from thermal runaway, the demand for heat from the secondary side of the exchanger must be consistently high. Under such optimal operating conditions, collection efficiencies exceeding 70% (up to ~35% electric, ~40% thermal for HCPVT) are anticipated. Net operating efficiencies may be substantially lower depending on how well a system is engineered to match the demands of the particular thermal application.

The maximum temperature of CPVT systems is typically too low (below 80-90 °C) to alone power a boiler for additional steam-based cogeneration of electricity. Such systems may be economical to power lower temperature applications having a constant high heat demand. The heat may be employed in district heating, water heating and air conditioning, desalination or process heat.

For applications having lower or intermittent heat demand, a system may be augmented with a switchable heat dump to the external environment in order to maintain reliable electrical output and safeguard cell life, despite the resulting reduction in net operating efficiency.

HCPVT active cooling enables the use of much higher power thermal–photovoltaic receiver units, generating typically 1–100 kilowatts electric, as compared to HCPV systems that mostly rely upon passive cooling of single ~20W cells. Such high-power receivers utilize dense arrays of cells mounted on a high-efficiency heat sink. Minimizing the number of individual receiver units is a simplification that should ultimately yield improvement in the overall balance of system costs, manufacturability, maintainability/upgradeability, and reliability.

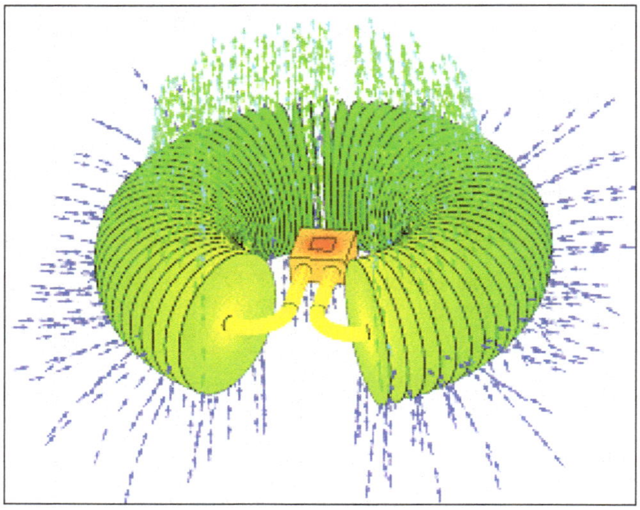

This 240 x 80 mm 1,000 suns CPV heat sink design thermal animation, was created using high resolution CFD analysis, and shows temperature contoured heat sink surface and flow trajectories as predicted.

Demonstration Projects

The economics of a mature CPVT industry is anticipated to be competitive, despite the large recent cost reductions and gradual efficiency improvements for conventional silicon PV (which can be installed alongside conventional CSP to provide for similar electrical+thermal generation capabilities). CPVT may currently be economical for niche markets having all of the following application characteristics:

- High solar direct normal incidence (DNI).
- Tight space constraints for placement of a solar collector array.
- High and constant demand for low-temperature (<80 °C) heat.
- High cost of grid electricity.
- Access to backup sources of power or cost-efficient storage (electrical and thermal).

Utilization of a power purchase agreement (PPA), government assistance programs, and innovative financing schemes are also helping potential manufacturers and users to mitigate the risks of early CPVT technology adoption.

CPVT equipment offerings ranging from low (LCPVT) to high (HCPVT) concentration are now being deployed by several startup ventures. As such, longer-term viability of the technical and/or business approach being pursued by any individual system provider is typically speculative. Notably, the minimum viable products of startups can vary widely in their attention to reliability engineering. Nevertheless, the following incomplete compilation is offered to assist with the identification of some early industry trends.

LCPVT systems at ~14x concentration using reflective trough concentrators, and receiver pipes clad with silicon cells having dense interconnects, have been assembled by Cogenra with a claimed 75% efficiency (~15-20% electric, 60% thermal). Several such systems are in operation for more than 5 years as of 2015, and similar systems are being produced by Absolicon and Idhelio at 10x and 50x concentration, respectively.

HCPVT offerings at over 700x concentration have more recently emerged, and may be classified into three power tiers. Third tier systems are distributed generators consisting of large arrays of ~20W single-cell receiver/collector units, similar to those previously pioneered by Amonix and SolFocus for HCPV. Second tier systems utilize localized dense-arrays of cells that produce 1-100 kW of electrical power output per receiver/generator unit. First tier systems exceed 100 kW of electrical output and are most aggressive in targeting the utility market.

Several HCPVT system providers are listed in the following table. Nearly all are early demonstration systems which have been in service for under 5 years as of 2015. Collected thermal power is typically 1.5x-2x the rated electrical power.

Provider	Country	Concentrator Type	Unit Size in kW_e	
			Generator	Receiver
		- Tier 1 -		
Raygen	Australia	Large Heliostat Array	250	250
		- Tier 2 -		
Airlight Energy/dsolar	Switzerland	Large Dish	12	12
Rehnu	United States	Large Dish	6.4	0.8
Solartron	Canada	Large Dish	20	20
Southwest Solar	United States	Large Dish	20	20
Sun Oyster	Germany	Large Trough + Lens	4.7	2.35
Zenith Solar/Suncore	Israel/China/ USA	Large Dish	4.5	2.25
		- Tier 3 -		
BSQ Solar	Spain	Small Lens Array	13,44	0.02
Silex Power	Malta	Small Dish Array	16	0.04
Solergy	Italy/USA	Small Lens Array	20	0.02

References

- "Hybrid power plants (wind- or solar-diesel)". TH-Energy.net – A platform for renewables & mining. Retrieved 12 May 2015
- Linear-concentrator-system-basics-concentrating-solar-power, solar, eere: energy.gov, Retrieved 13 February, 2019

- Kunal K. Shah, Aishwarya S. Mundada, Joshua M. Pearce. Performance of U.S. hybrid distributed energy systems: Solar photovoltaic, battery and combined heat and power. Energy Conversion and Management 105, pp. 71–80 (2015). DOI: 10.1016/j.enconman.2015.07.048
- Dishengine-system-concentrating-solar-power-basics, solar, eere: energy.gov, Retrieved 14 March, 2019
- "Photovoltaic concentrators (CPV) - Performance testing - Part 1: Standard conditions". Www.iec.ch. Archived from the original on 2019-01-24. Retrieved 2019-01-20
- López, Antonio Luque; Andreev, Viacheslav M. (2007). Past Experiences and New Challenges of PV Concentrators, G Sala and A Luque, Springer Series in Optical Sciences 130, 1, (2007). Springer Series in Optical Sciences. 130. Doi:10.1007/978-3-540-68798-6. ISBN 978-3-540-68796-2
- Photovoltaics, initiatives: seia.org, Retrieved 12 January, 2019
- Ahmad Mojiri, Robert A. Taylor, Elizabeth Thomsen, Gary Rosengarten, Spectral beam splitting for efficient conversion of solar energy — A review. In: Renewable and Sustainable Energy Reviews 28, December 2013, Pages 654–663, doi:10.1016/j.rser.2013.08.026
- Power-tower-system-concentrating-solar-power-basics, solar, eere: energy.gov, Retrieved 15 April, 2019

Solar Cell

CHAPTER 3

The electrical device which is used for the conversion of light energy into electricity by the photovoltaic effect is known as solar cell. Plasmonic solar cells, thin film solar cells, perovskite solar cells, multi-junction solar cells, dye-sensitized solar cells, organic solar cells, hybrid solar cells and quantum dot solar cells are the various types of solar cells. The topics elaborated in this chapter will help in gaining a better perspective about these types of solar cells.

Solar cell, also called photovoltaic cell, is any device that directly converts the energy of light into electrical energy through the photovoltaic effect. The overwhelming majority of solar cells are fabricated from silicon—with increasing efficiency and lowering cost as the materials range from amorphous (noncrystalline) to polycrystalline to crystalline (single crystal) silicon forms. Unlike batteries or fuel cells, solar cells do not utilize chemical reactions or require fuel to produce electric power, and, unlike electric generators, they do not have any moving parts.

Solar cells can be arranged into large groupings called arrays. These arrays, composed of many thousands of individual cells, can function as central electric power stations, converting sunlight into electrical energy for distribution to industrial, commercial, and residential users. Solar cells in much smaller configurations, commonly referred to as solar cell panels or simply solar panels, have been installed by homeowners on their rooftops to replace or augment their conventional electric supply. Solar cell panels also are used to provide electric power in many remote terrestrial locations where conventional electric power sources are either unavailable or prohibitively expensive to install. Because they have no moving parts that could need maintenance or fuels that would require replenishment, solar cells provide power for most space installations, from communications and weather satellites to space stations. (Solar power is insufficient for space probes sent to the outer planets of the solar system or into interstellar space, however, because of the diffusion of radiant energy with distance from the Sun.) Solar cells have also been used in consumer products, such as electronic toys, handheld calculators, and portable radios. Solar cells used in devices of this kind may utilize artificial light (e.g., from incandescent and fluorescent lamps) as well as sunlight.

International Space Station.

The International Space Station (ISS) was built in sections beginning in 1998. By December 2000 the major elements of the partially completed station included the American-built connecting node Unity and two Russian-built units—Zarya, a power module, and Zvezda, the initial living quarters. A Russian spacecraft, which carried up the station's first three-person crew, is docked at the end of Zvezda. The photograph was taken from the space shuttle Endeavour.

While total photovoltaic energy production is minuscule, it is likely to increase as fossil fuel resources shrink. In fact, calculations based on the world's projected energy consumption by 2030 suggest that global energy demands would be fulfilled by solar panels operating at 20 percent efficiency and covering only about 496,805 square km (191,817 square miles) of Earth's surface. The material requirements would be enormous but feasible, as silicon is the second most abundant element in Earth's crust. These factors have led solar proponents to envision a future "solar economy" in which practically all of humanity's energy requirements are satisfied by cheap, clean, renewable sunlight.

Solar Cell Structure and Operation

Solar cells, whether used in a central power station, a satellite, or a calculator, have the same basic structure. Light enters the device through an optical coating, or antireflection layer, that minimizes the loss of light by reflection; it effectively traps the light falling on the solar cell by promoting its transmission to the energy-conversion layers below. The antireflection layer is typically an oxide of silicon, tantalum, or titanium that is formed on the cell surface by spin-coating or a vacuum deposition technique.

Solar energy; solar cell: A solar energy plant produces megawatts of electricity. Voltage is generated by solar cells made from specially treated semiconductor materials, such as silicon.

The three energy-conversion layers below the antireflection layer are the top junction layer, the absorber layer, which constitutes the core of the device, and the back junction layer. Two additional electrical contact layers are needed to carry the electric current out to an external load and back into the cell, thus completing an electric circuit. The electrical contact layer on the face of the cell where light enters is generally present in some grid pattern and is composed of a good conductor such as a metal. Since metal blocks light, the grid lines are as thin and widely spaced as is possible without impairing collection of the current produced by the cell. The back electrical contact layer has no such diametrically opposed restrictions. It need simply function as an electrical contact and thus covers the entire back surface of the cell structure. Because the back layer also must be a very good electrical conductor, it is always made of metal.

Since most of the energy in sunlight and artificial light is in the visible range of electromagnetic radiation, a solar cell absorber should be efficient in absorbing radiation at those wavelengths. Materials that strongly absorb visible radiation belong to a class of substances known as semiconductors. Semiconductors in thicknesses of about one-hundredth of a centimetre or less can absorb all incident visible light; since the junction-forming and contact layers are much thinner, the thickness of a solar cell is essentially that of the absorber. Examples of semiconductor materials employed in solar cells include silicon, gallium arsenide, indium phosphide, and copper indium selenide.

When light falls on a solar cell, electrons in the absorber layer are excited from a lower-energy "ground state," in which they are bound to specific atoms in the solid, to a higher "excited state," in which they can move through the solid. In the absence of the junction-forming layers, these "free" electrons are in random motion, and so there can be no oriented direct current. The addition of junction-forming layers, however, induces a built-in electric field that produces the photovoltaic effect. In effect, the electric field gives a collective motion to the electrons that flow past the electrical contact layers into an external circuit where they can do useful work.

The materials used for the two junction-forming layers must be dissimilar to the absorber in order to produce the built-in electric field and to carry the electric current. Hence, these may be different semiconductors (or the same semiconductor with different types of conduction), or they may be a metal and a semiconductor. The materials used to construct the various layers of solar cells are essentially the same as those used to produce the diodes and transistors of solid-state electronics and microelectronics. Solar cells and microelectronic devices share the same basic technology. In solar cell fabrication, however, one seeks to construct a large-area device because the power produced is proportional to the illuminated area. In microelectronics the goal is, of course, to construct electronic components of ever smaller dimensions in order to increase their density and operating speed within semiconductor chips, or integrated circuits.

The photovoltaic process bears certain similarities to photosynthesis, the process by which the energy in light is converted into chemical energy in plants. Since solar cells obviously cannot produce electric power in the dark, part of the energy they develop under light is stored, in many applications, for use when light is not available. One common means of storing this electrical energy is by charging electrochemical storage batteries. This sequence of converting the energy in light into the energy of excited electrons and then into stored chemical energy is strikingly similar to the process of photosynthesis.

Solar Panel Design

Most solar cells are a few square centimetres in area and protected from the environment by a thin coating of glass or transparent plastic. Because a typical 10 cm × 10 cm (4 inch × 4 inch) solar cell generates only about two watts of electrical power (15 to 20 percent of the energy of light incident on their surface), cells are usually combined in series to boost the voltage or in parallel to increase the current. A solar, or photovoltaic (PV), module generally consists of 36 interconnected cells laminated to glass within an aluminum frame. In turn, one or more of these modules may be wired and framed together to form a solar panel. Solar panels are slightly less efficient at energy conversion per surface area than individual cells, because of inevitable inactive areas in the assembly and cell-to-cell variations in performance. The back of each solar panel is equipped with standardized sockets so that its output can be combined with other solar panels to form a solar array. A complete photovoltaic system may consist of many solar panels, a power system for accommodating different electrical loads, an external circuit, and storage batteries. Photovoltaic systems are broadly classifiable as either stand-alone or grid-connected systems.

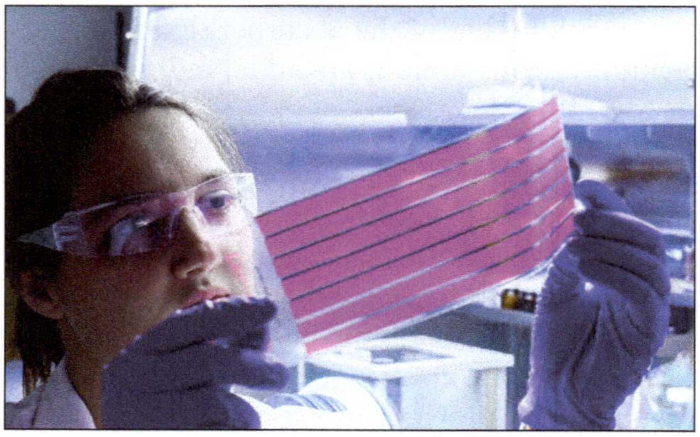

Solar cell: A scientist examines a sheet of polymer solar cells, which are more lightweight, more flexible, and cheaper than traditional silicon solar cells.

Stand-alone systems contain a solar array and a bank of batteries directly wired to an application or load circuit. A battery system is essential to compensate for the absence of any electrical output from the cells at night or in overcast conditions; this adds considerably to the overall cost. Each battery stores direct current (DC) electricity at a fixed voltage determined by the panel specifications, although load requirements may differ. DC-to-DC converters are used to provide the voltage levels demanded by DC loads, and DC-to-AC inverters supply power to alternating current (AC) loads. Stand-alone systems are ideally suited for remote installations where linking to a central power station is prohibitively expensive. Examples include pumping water for feedstock and providing electric power to lighthouses, telecommunications repeater stations, and mountain lodges.

Grid-connected systems integrate solar arrays with public utility power grids in two ways. One-way systems are used by utilities to supplement power grids during midday peak usage. Bidirectional systems are used by companies and individuals to supply some or all of their power needs, with any excess power fed back into a utility power grid. A major advantage of grid-connected systems is that no storage batteries are needed. The corresponding reduction in capital and maintenance costs is offset, however, by the increased complexity of the system. Inverters and additional protective gear are needed to interface low-voltage DC output from the solar array with a high-voltage AC power grid. Additionally, rate structures for reverse metering are necessary when residential and industrial solar systems feed energy back into a utility grid.

The simplest deployment of solar panels is on a tilted support frame or rack known as a fixed mount. For maximum efficiency, a fixed mount should face south in the Northern Hemisphere or north in the Southern Hemisphere, and it should have a tilt angle from horizontal of about 15 degrees less than the local latitude in summer and 25 degrees more than the local latitude in winter. More complicated deployments involve motor-driven tracking systems that continually reorient the panels to follow the daily and seasonal movements of the Sun. Such systems are justified only for large-scale utility generation using high-efficiency concentrator solar cells with lenses or parabolic mirrors that can intensify solar radiation a hundredfold or more.

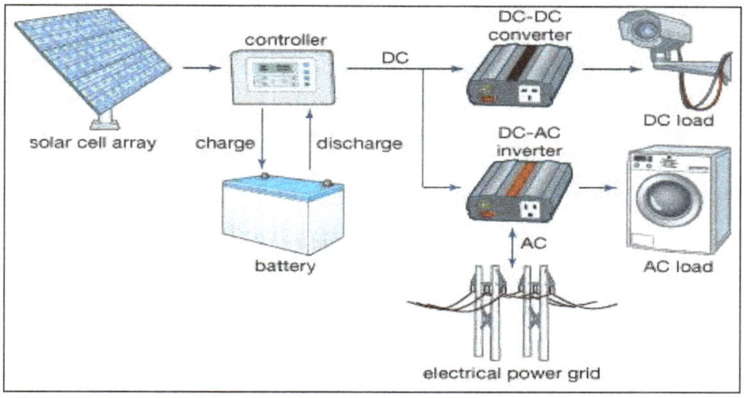

A grid-connected solar cell system.

Although sunlight is free, the cost of materials and available space must be considered in designing a solar system; less-efficient solar panels imply more panels, occupying more space, in order to produce the same amount of electricity. Compromises between cost of materials and efficiency are particularly evident for space-based solar systems. Panels used on satellites have to be extra-rugged, reliable, and resistant to radiation damage encountered in Earth's upper atmosphere. In addition, minimizing the liftoff weight of these panels is more critical than fabrication costs. Another factor in solar panel design is the ability to fabricate cells in "thin-film" form on a variety of substrates, such as glass, ceramic, and plastic, for more flexible deployment. Amorphous silicon is very attractive from this viewpoint. In particular, amorphous silicon-coated roof tiles and other photovoltaic materials have been introduced in architectural design and for recreational vehicles, boats, and automobiles.

Thin-film solar cell: Thin-film solar cells, such as those used in solar panels, convert light energy into electrical energy.

Development of Solar Cells

The development of solar cell technology stems from the work of French physicist Antoine-César Becquerel in 1839. Becquerel discovered the photovoltaic effect while experimenting with a solid electrode in an electrolyte solution; he observed that voltage developed when light fell upon the electrode. About 50 years later, Charles Fritts constructed the first true solar cells using junctions formed by coating the semiconductor selenium with an ultrathin, nearly transparent layer of gold. Fritts's devices were very inefficient converters of energy; they transformed less than 1 percent of absorbed light energy into electrical energy. Though inefficient by today's standards, these early solar cells fostered among some a vision of abundant, clean power. In 1891 R. Appleyard wrote of:

> "The blessed vision of the Sun, no longer pouring his energies unrequited into space, but by means of photo-electric cells, these powers gathered into electrical storehouses to the total extinction of steam engines, and the utter repression of smoke."

By 1927 another metal-semiconductor-junction solar cell, in this case made of copper and the semiconductor copper oxide, had been demonstrated. By the 1930s both the selenium cell and the copper oxide cell were being employed in light-sensitive devices, such as photometers, for use in photography. These early solar cells, however, still had energy-conversion efficiencies of less than 1 percent. This impasse was finally overcome with the development of the silicon solar cell by Russell Ohl in 1941. Thirteen years later, aided by the rapid commercialization of silicon technology needed to fabricate the transistor, three other American researchers—Gerald Pearson, Daryl Chapin, and Calvin Fuller—demonstrated a silicon solar cell capable of a 6 percent energy-conversion efficiency when used in direct sunlight. By the late 1980s silicon cells, as well as cells made of gallium arsenide, with efficiencies of more than 20 percent had been fabricated. In 1989 a concentrator solar cell in which sunlight was concentrated onto the cell surface by means of lenses achieved an efficiency of 37 percent owing to the increased intensity of the collected energy. By connecting cells of different semiconductors optically and electrically in series, even higher efficiencies are possible, but at increased cost and added complexity. In general, solar cells of widely varying efficiencies and cost are now available.

PLASMONIC SOLAR CELL

A plasmonic-enhanced solar cell, is a type of solar cell (including thin-film, crystalline silicon, amorphous silicon, and other types of cells) that converts light into electricity with the assistance of plasmons, but where the photovoltaic effect occurs in another material.

A direct plasmonic solar cell is a solar cell that converts light into electricity using plasmons as the active, photovoltaic material.

The thickness varies from that of traditional silicon PV, to less than 2 μm thick and theoretically could be as thin as 100 nm. They can use substrates which are cheaper than silicon, such as glass, plastic or steel. One of the challenges for thin film solar cells is that they do not absorb as much light as thicker solar cells made with materials with the same absorption coefficient. Methods for light trapping are important for thin film solar cells. Plasmonic-enhanced cells improve absorption by scattering light using metal nano-particles excited at their surface plasmon resonance. Interestingly, plasmonic core-shell nanoparticles located in the front of the thin film solar cells can aid weak absorption of Si solar cells in the near-infrared region—the fraction of light scattered into the substrate and the maximum optical path length enhancement can be as high as 0.999 and 3133. Incoming light at the plasmon resonance frequency induces electron oscillations at the surface of the nanoparticles. The oscillation electrons can then be captured by a conductive layer producing an electrical current. The voltage produced is dependent on the bandgap of the conductive layer and the potential of the electrolyte in contact with the nanoparticles. There is still considerable research necessary to enable the technology to reach its full potential and commercialization of plasmonic-enhanced solar cells.

Design

The design for plasmonic-enhanced solar cells varies depending on the method being used to trap and scatter light across the surface and through the material.

Nanoparticle Cells

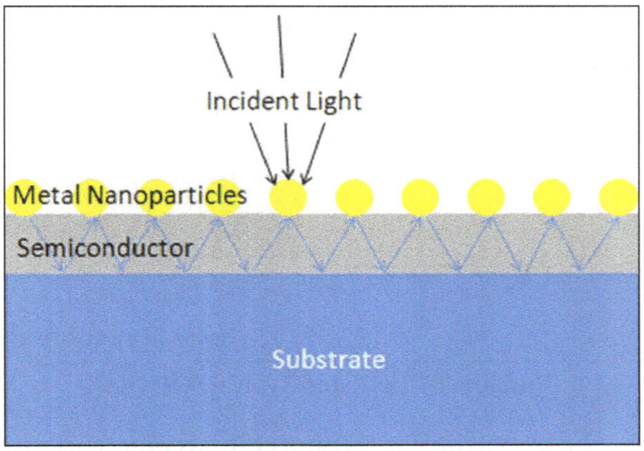

PSC using metal nano-particles.

A common design is to deposit metal nano-particles on the top surface of the surface of the solar cell. When light hits these metal nano-particles at their surface plasmon resonance, the light is scattered in many different directions. This allows light to travel along the solar cell and bounce between the substrate and the nano-particles enabling the solar cell to absorb more light. The concentrated near field intensity induced by localized surface plasmon of the metal nanoparticles will promote the optical absorption of semiconductors. Recently, the plasmonic asymmetric modes of nano particles have found to favor the broadband optical absorption and promote the electrical properties of solar cells. The simultaneously plasmon-optical and plasmon-electrical effects of nanoparticles reveal a promising feature of nanoparticle plasmon.

Recently, the core (metal)-shell (dielectric) nanoparticle has demonstrated a zero backward scattering with enhanced forward scattering on Si substrate when surface plasmon is located in front of a solar cell. The core-shell nanoparticles can support simultaneously both electric and magnetic resonances, demonstrating entirely new properties when compared with bare metallic nanoparticles if the resonances are properly engineered.

Metal Film Cells

Other methods utilizing surface plasmons for harvesting solar energy are available. One other type of structure is to have a thin film of silicon and a thin layer of metal deposited on the lower surface. The light will travel through the silicon and generate surface plasmons on the interface of the silicon and metal. This generates electric fields inside of the silicon since electric fields do not travel very far into metals. If the electric field is strong enough, electrons can be moved and collected to produce a photocurrent. The thin film of metal in this design must have nanometer sized grooves which act as waveguides for the incoming light in order to excite as many photons in the silicon thin film as possible.

Principles

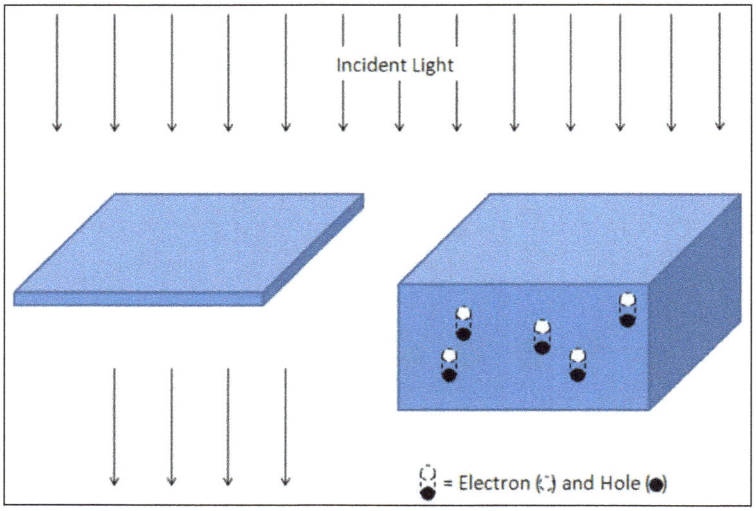

Thin film SC (left) and Typical SC (right).

When a photon is excited in the substrate of a solar cell, an electron and hole are separated. Once the electrons and holes are separated, they will want to recombine since they are of opposite charge. If the electrons can be collected prior to this happening they can be used as a current for an external circuit. Designing the thickness of a solar cell is always a trade-off between minimizing this recombination (thinner layers) and absorbing more photons (thicker layer).

Nano-particles

Scattering and Absorption

The basic principles for the functioning of plasmonic-enhanced solar cells include scattering and absorption of light due to the deposition of metal nano-particles. Silicon does not absorb light very well. For this reason, more light needs to be scattered across the surface in order to increase the absorption. It has been found that metal nano-particles help to scatter the incoming light across the surface of the silicon substrate. The equations that govern the scattering and absorption of light can be shown as:

$$C_{scat} = \frac{1}{6\pi}\left(\frac{2\pi}{\lambda}\right)^4 |\alpha|^2$$

This shows the scattering of light for particles which have diameters below the wavelength of light.

$$C_{abs} = \frac{2\pi}{\lambda}\operatorname{Im}[\alpha]$$

This shows the absorption for a point dipole model.

$$\alpha = 3V\left[\frac{\epsilon_p / \epsilon_m - 1}{\epsilon_p / \epsilon_m + 2}\right]$$

This is the polarizability of the particle. V is the particle volume. ϵ_p is the dielectric function of the particle. ϵ_m is the dielectric function of the embedding medium. When $\epsilon_p = -2\epsilon_m$ the polarizability of the particle becomes large. This polarizability value is known as the surface plasmon resonance. The dielectric function for metals with low absorption can be defined as:

$$\epsilon = 1 - \frac{\omega_p^2}{\omega^2 + i\gamma\omega}$$

In the previous equation, ω_p is the bulk plasma frequency. This is defined as:

$$\omega_p^2 = Ne^2 / m\epsilon_0$$

N is the density of free electrons, e is the electronic charge and m is the effective mass of an electron. ϵ_0 is the dielectric constant of free space. The equation for the surface plasmon resonance in free space can therefore be represented by:

$$\alpha = 3V\frac{\omega_p^2}{\omega_p^2 - 3\omega^2 - i\gamma\omega}$$

Many of the plasmonic solar cells use nano-particles to enhance the scattering of light. These nano-particles take the shape of spheres, and therefore the surface plasmon resonance frequency for spheres is desirable. By solving the previous equations, the surface plasmon resonance frequency for a sphere in free space can be shown as:

$$\omega_{sp} = \sqrt{3}\omega_p$$

As an example, at the surface plasmon resonance for a silver nanoparticle, the scattering cross-section is about 10x the cross-section of the nanoparticle. The goal of the nano-particles is to trap light on the surface of the SC. The absorption of light is not important for the nanoparticle, rather, it is important for the SC. One would think that if the nanoparticle is increased in size, then the scattering cross-section becomes larger. This is true, however, when compared with the size of the nanoparticle, the ratio ($CS_{scat} / CS_{particle}$) is reduced. Particles with a large scattering cross section tend to have a broader plasmon resonance range.

Wavelength Dependence

Surface plasmon resonance mainly depends on the density of free electrons in the particle. The order of densities of electrons for different metals is shown below along with the type of light which corresponds to the resonance.

- Aluminum - Ultra-violet

- Silver - Ultra-violet
- Gold - Visible
- Copper - Visible

If the dielectric constant for the embedding medium is varied, the resonant frequency can be shifted. Higher indexes of refraction will lead to a longer wavelength frequency.

Light Trapping

The metal nano-particles are deposited at a distance from the substrate in order to trap the light between the substrate and the particles. The particles are embedded in a material on top of the substrate. The material is typically a dielectric, such as silicon or silicon nitride. When performing experiment and simulations on the amount of light scattered into the substrate due to the distance between the particle and substrate, air is used as the embedding material as a reference. It has been found that the amount of light radiated into the substrate decreases with distance from the substrate. This means that nano-particles on the surface are desirable for radiating light into the substrate, but if there is no distance between the particle and substrate, then the light is not trapped and more light escapes.

The surface plasmons are the excitations of the conduction electrons at the interface of metal and the dielectric. Metallic nano-particles can be used to couple and trap freely propagating plane waves into the semiconductor thin film layer. Light can be folded into the absorbing layer to increase the absorption. The localized surface plasmons in metal nano-particles and the surface plasmon polaritons at the interface of metal and semiconductor are of interest in the current research. In recent reported papers, the shape and size of the metal nano-particles are key factors to determine the incoupling efficiency. The smaller particles have larger incoupling efficiency due to the enhanced near-field coupling. However, very small particles suffer from large ohmic losses.

Recently, the plasmonic asymmetric modes of nano particles have found to favor the broadband optical absorption and promote the electrical properties of solar cells. The simultaneously plasmon-optical and plasmon-electrical effects of nanoparticles reveal a promising feature of nanoparticle plasmon.

Metal Film

As light is incident upon the surface of the metal film, it excites surface plasmons. The surface plasmon frequency is specific for the material, but through the use of gratings on the surface of the film, different frequencies can be obtained. The surface plasmons are also preserved through the use of waveguides as they make the surface plasmons easier to travel on the surface and the losses due to resistance and radiation are minimized. The electric field generated by the surface plasmons influences the electrons to travel toward the collecting substrate.

Materials

First Generation	Second Generation	Third Generation
Single-crystal silicon	CuInSe2	Gallium Indium Phosphide

Multicrystalline silicon	amorphous silicon	Gallium Indium Arsenide
Polycrystalline silicon	thin film crystalline Si	Germanium

Applications

The applications for plasmonic-enhanced solar cells are endless. The need for cheaper and more efficient solar cells is huge. In order for solar cells to be considered cost effective, they need to provide energy for a smaller price than that of traditional power sources such as coal and gasoline. The movement toward a more green world has helped to spark research in the area of plasmonic-enhanced solar cells. Currently, solar cells cannot exceed efficiencies of about 30% (First Generation). With new technologies (Third Generation), efficiencies of up to 40-60% can be expected. With a reduction of materials through the use of thin film technology (Second Generation), prices can be driven lower.

Certain applications for plasmonic-enhanced solar cells would be for space exploration vehicles. A main contribution for this would be the reduced weight of the solar cells. An external fuel source would also not be needed if enough power could be generated from the solar cells. This would drastically help to reduce the weight as well.

Solar cells have a great potential to help rural electrification. An estimated two million villages near the equator have limited access to electricity and fossil fuels and that approximately 25% of people in the world do not have access to electricity. When the cost of extending power grids, running rural electricity and using diesel generators is compared with the cost of solar cells, many times the solar cells win. If the efficiency and cost of the current solar cell technology is decreased even further, then many rural communities and villages around the world could obtain electricity when current methods are out of the question. Specific applications for rural communities would be water pumping systems, residential electric supply and street lights. A particularly interesting application would be for health systems in countries where motorized vehicles are not overly abundant. Solar cells could be used to provide the power to refrigerate medications in coolers during transport.

Solar cells could also provide power to lighthouses, buoys, or even battleships out in the ocean. Industrial companies could use them to power telecommunications systems or monitoring and control systems along pipelines or other system.

If the solar cells could be produced on a large scale and be cost effective then entire power stations could be built in order to provide power to the electrical grids. With a reduction in size, they could be implemented on both commercial and residential buildings with a much smaller footprint. They might not even seem like an eyesore.

Other areas are in hybrid systems. The solar cells could help to power high consumption devices such as automobiles in order to reduce the amount of fossil fuels used and to help improve the environmental conditions of the earth.

In consumer electronics devices, solar cells could be used to replace batteries for low power electronics. This would save everyone a lot of money and it would also help to reduce the amount of waste going into landfills.

Recent Advancements

Choice of Plasmonic Metal Nano-particles

Proper choice of plasmatic metal nano-particles is crucial for the maximum light absorption in the active layer. Front surface located nano-particles Ag and Au are the most widely used materials due to their surface plasmon resonances located in the visible range and therefore interact more strongly with the peak solar intensity. However, such noble metal nano-particles always introduce reduced light coupling into Si at the short wavelengths below the surface plasmon resonance due to the detrimental Fano effect, i.e. the destructive interference between the scattered and unscattered light. Moreover, the noble metal nano-particles are impractical to implement for large-scale solar cell manufacture due to their high cost and scarcity in the earth's crust. Recently, Zhang et al. have demonstrated the low cost and earth abundant materials Al nano-particles to be able to outperform the widely used Ag and Au nano-particles. Al nano-particles, with their surface plasmon resonances located in the UV region below the desired solar spectrum edge at 300 nm, can avoid the reduction and introduce extra enhancement in the shorter wavelength range.

Shape Choice of Nano-particles

Shape

- Nanosphere
- Nanostar
- Core-shell nanoparticle
- Nanodisk
- Nanocavity
- Nanovoid
- Nucleated nanoparticle
- Nanocage
- Core-shell nanoparticle

Light Trapping

Being able to concentrate and scatter light across the surface of the plasmonic-enhanced solar cell will help to increase efficiencies. Recently, research at Sandia National Laboratories has discovered a photonic waveguide which collects light at a certain wavelength and traps it within the structure. This new structure can contain 95% of the light that enters it compared to 30% for other traditional waveguides. It can also direct the light within one wavelength which is ten times greater than traditional waveguides. The wavelength this device captures can be selected by changing the structure of the lattice which comprises the structure. If this structure is used to trap light and keep it in the structure until the solar cell can absorb it, the efficiency of the solar cell could be increased dramatically.

Absorption

Another recent advancement in plasmonic-enhanced solar cells is using other methods to aid in the absorption of light. One way being researched is the use of metal wires on top of the substrate to scatter the light. This would help by utilizing a larger area of the surface of the solar cell for light scattering and absorption. The danger in using lines instead of dots would be creating a reflective layer which would reject light from the system. This is very undesirable for solar cells. This would be very similar to the thin metal film approach, but it also utilizes the scattering effect of the nano-particles. Yue, et al. used a type of new materials, called topological insulators, to increase the absorption of ultrathin a-Si solar cells. The topological insulator nanostructure has intrinsically core-shell configuration. The core is dielectric and has ultrahigh refractive index. The shell is metallic and support surface plasmon resonances. Through integrating the nanocone arrays into a-Si thin film solar cells, up to 15% enhancement of light absorption was predicted in the ultraviolet and visible ranges.

Third Generation

The goal of third generation solar cells is to increase the efficiency using second generation solar cells (thin film) and using materials that are found abundantly on earth. This has also been a goal of the thin film solar cells. With the use of common and safe materials, third generation solar cells should be able to be manufactured in mass quantities further reducing the costs. The initial costs would be high in order to produce the manufacturing processes, but after that they should be cheap. The way third generation solar cells will be able to improve efficiency is to absorb a wider range of frequencies. The current thin film technology has been limited to one frequency due to the use of single band gap devices.

Multiple Energy Levels

The idea for multiple energy level solar cells is to basically stack thin film solar cells on top of each other. Each thin film solar cell would have a different band gap which means that if part of the solar spectrum was not absorbed by the first cell then the one just below would be able to absorb part of the spectrum. These can be stacked and an optimal band gap can be used for each cell in order to produce the maximum amount of power. Options for how each cell is connected are available, such as serial or parallel. The serial connection is desired because the output of the solar cell would just be two leads.

The lattice structure in each of the thin film cells needs to be the same. If it is not then there will be losses. The processes used for depositing the layers are complex. They include Molecular Beam Epitaxy and Metal Organic Vapour Phase Epitaxy. The current efficiency record is made with this process but doesn't have exact matching lattice constants. The losses due to this are not as effective because the differences in lattices allows for more optimal band gap material for the first two cells. This type of cell is expected to be able to be 50% efficient.

Lower quality materials that use cheaper deposition processes are being researched as well. These devices are not as efficient, but the price, size and power combined allow them to be just as cost effective. Since the processes are simpler and the materials are more readily available, the mass production of these devices is more economical.

Hot Carrier Cells

A problem with solar cells is that the high energy photons that hit the surface are converted to heat. This is a loss for the cell because the incoming photons are not converted into usable energy. The idea behind the hot carrier cell is to utilize some of that incoming energy which is converted to heat. If the electrons and holes can be collected while hot, a higher voltage can be obtained from the cell. The problem with doing this is that the contacts which collect the electrons and holes will cool the material. Thus far, keeping the contacts from cooling the cell has been theoretical. Another way of improving the efficiency of the solar cell using the heat generated is to have a cell which allows lower energy photons to excite electron and hole pairs. This requires a small bandgap. Using a selective contact, the lower energy electrons and holes can be collected while allowing the higher energy ones to continue moving through the cell. The selective contacts are made using a double barrier resonant tunneling structure. The carriers are cooled which they scatter with phonons. If a material with a large bandgap of phonons then the carriers will carry more of the heat to the contact and it won't be lost in the lattice structure. One material which has a large bandgap of phonons is indium nitride. The hot carrier cells are in their infancy but are beginning to move toward the experimental stage.

Plasmonic-electrical Solar Cells

Having unique features of tunable resonances and unprecedented near-field enhancement, plasmon is an enabling technique for light management. Recently, performances of thin-film solar cells have been pronouncedly improved by introducing metallic nanostructures. The improvements are mainly attributed to the plasmonic-optical effects for manipulating light propagation, absorption, and scattering. The plasmonic-optical effects could: (1) boost optical absorption of active materials; (2) spatially redistribute light absorption at the active layer due to the localized near-field enhancement around metallic nanostructures. Except for the plasmonic-optical effects, the effects of plasmonically modified recombination, transport and collection of photocarriers (electrons and holes), hereafter named plasmonic-electrical effects, have been proposed by Sha, etal. For boosting device performance, they conceived a general design rule, tailored to arbitrary electron to hole mobility ratio, to decide the transport paths of photocarriers. The design rule suggests that electron to hole transport length ratio should be balanced with electron to hole mobility ratio. In other words, the transport time of electrons and holes (from initial generation sites to corresponding electrodes) should be the same. The general design rule can be realized by spatially redistributing light absorption at the active layer of devices (with the plasmonic-electrical effect). They also demonstrated the breaking of space charge limit in plasmonic-electrical organic solar cell. Recently, the plasmonic asymmetric modes of nano particles have found to favor the broadband optical absorption and promote the electrical properties of solar cells. The simultaneously plasmon-optical and plasmon-electrical effects of nanoparticles reveal a promising feature of nanoparticle plasmon.

Ultra-thin Plasmonic Wafer Solar Cells

Reducing the silicon wafer thickness at a minimized efficiency loss represents a mainstream trend in increasing the cost-effectiveness of wafer-based solar cells. Recently, Zhang et al. have demonstrated that, using the advanced light trapping strategy with a properly designed nano-particle architecture, the wafer thickness can be dramatically reduced to only around 1/10 of the current thickness (180 μm) without any solar cell efficiency loss at 18.2%. Nano-particle integrated ultra-thin solar cells with only 3% of the current wafer thickness can potentially achieve 15.3% efficiency combining the absorption enhancement with the benefit of thinner wafer induced open circuit voltage increase. This represents a 97% material saving with only 15% relative efficiency loss. These results demonstrate the feasibility and prospect of achieving high-efficiency ultra-thin silicon wafer cells with plasmonic light trapping.

THIN FILM SOLAR CELL

Thin-film solar cell is a type of device that is designed to convert light energy into electrical energy (through the photovoltaic effect) and is composed of micron-thick photon-absorbing material layers deposited over a flexible substrate. Thin-film solar cells were originally introduced in the 1970s by researchers at the Institute of Energy Conversion at the University of Delaware in the United States. The technology continuously improved so that in the early 21st century the global thin-film photovoltaic market was growing at an unprecedented rate and was forecast to continue to grow. Several types of thin-film solar cells are widely used because of their relatively low cost and their efficiency in producing electricity.

Types of Thin-film Solar Cells

Cadmium telluride thin-film solar cells are the most common type available. They are less expensive than the more standard silicon thin-film cells. Cadmium telluride thin-films have a peak recorded efficiency of more than 18 percent (the percentage of photons hitting the surface of the cell that are transformed into an electric current). By 2014 cadmium telluride thin-film technologies had the smallest carbon footprint and quickest payback time of any thin-film solar cell technology on the market (payback time being the time it takes for the solar panel's electricity generation to cover the cost of purchase and installation).

Copper indium gallium selenide (CIGS) is another type of semiconductor used to manufacture thin-film solar cells. CIGS thin-film solar cells have reached 20 percent efficiency in laboratory settings and 14 percent efficiency in the field, making CIGS a leader among alternative cell materials and a promising semiconducting material in thin-film technologies. CIGS cells traditionally have been more costly than other types of cells on the market, and for that reason they are not widely used.

Gallium arsenide (GaAs) thin-film solar cells have reached nearly 30 percent efficiency in laboratory environments, but they are very expensive to manufacture. Cost has been a major factor in limiting the market for GaAs solar cells; their main use has been for spacecraft and satellites.

Amorphous silicon thin-film cells are the oldest and most mature type of thin-film. They are made of noncrystalline silicon, unlike typical solar-cell wafers. Amorphous silicon is cheaper to manufacture than crystalline silicon and most other semiconducting materials. Amorphous silicon is also popular because it is abundant, nontoxic, and relatively inexpensive. However, the average efficiency is very low, 10 percent.

Applications of Thin-film Solar Cells

Applications of thin-film solar cells began in the 1980s with small strips that were used for calculators and watches. Throughout the early 21st century the potential for thin-film applications increased greatly, because of their flexibility, which facilitates their installation on curved surfaces as well as their use in building-integrated photovoltaics.

However, standard and rigid photovoltaics, such as classic crystalline silicon panels, outperform thin-films in efficiency. With the exception of cadmium telluride thin-films, nonflexible photovoltaic cells have faster payback times, and their construction is more durable, which has advantages in many applications. The advantages of both types of solar cells raise two questions: What does the consumer or client prefer? and Which type will perform best for a particular application?

As thin-film solar cells continue to improve in efficiency, it is predicted that they could overtake the classic inflexible photovoltaic technologies that have been in use since the mid-20th century. Sheets of thin-films may be used to generate electricity increasingly in places where other photovoltaic cells cannot be used, such as on curved surfaces on buildings or cars or even on clothing to charge handheld devices. Such uses could help to achieve a sustainable energy future.

PEROVSKITE SOLAR CELL

A perovskite solar cell (PSC) is a type of solar cell which includes a perovskite structured compound, most commonly a hybrid organic-inorganic lead or tin halide-based material, as the light-harvesting active layer. Perovskite materials, such as methylammonium lead halides and all-inorganic cesium lead halide, are cheap to produce and simple to manufacture.

Solar cell efficiencies of devices using these materials have increased from 3.8% in 2009 to 24.2% in 2019 in single-junction architectures, and, in silicon-based tandem cells, to 28.0%, exceeding the maximum efficiency achieved in single-junction silicon solar cells. Perovskite solar cells are therefore the fastest-advancing solar technology to date. With the potential of achieving even higher efficiencies and very low production costs, perovskite solar cells have become commercially attractive.

Advantages

Metal halide perovskites possess unique features that make them useful for solar cell applications. The raw materials used, and the possible fabrication methods (such as various printing techniques) are both low cost. Their high absorption coefficient enables ultrathin films of around 500 nm to absorb the complete visible solar spectrum. These features combined result in the possibility to create low cost, high efficiency, thin, lightweight and flexible solar modules.

Materials

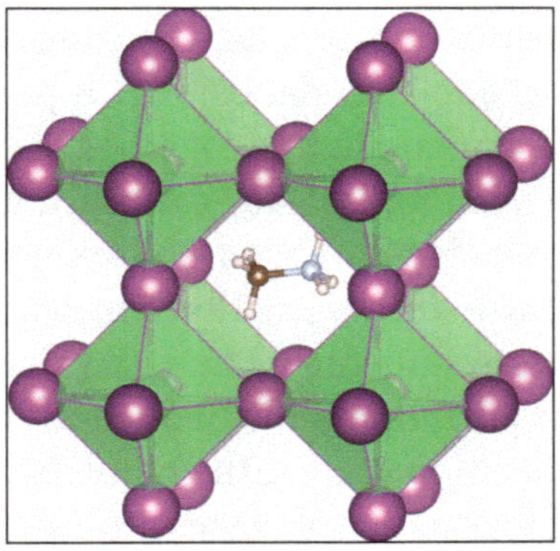

Crystal structure of $CH_3NH_3PbX_3$ perovskites (X=I, Br and/or Cl). The methylammonium cation ($CH_3NH_3^+$) is surrounded by PbX_6 octahedra.

The name 'perovskite solar cell' is derived from the ABX_3 crystal structure of the absorber materials, which is referred to as perovskite structure. The most commonly studied perovskite absorber is methylammonium lead trihalide ($CH_3NH_3PbX_3$, where X is a halogen atom such as iodine, bromine or chlorine), with an optical bandgap between 1.5 and 2.3 eV depending on halide content. Formamidinum lead trihalide ($H_2NCHNH_2PbX_3$) has also shown promise, with bandgaps between 1.5 and 2.2 eV. The minimum bandgap is closer to the optimal for a single-junction cell than methylammonium lead trihalide, so it should be capable of higher efficiencies. The first use of perovskite in a solid state solar cell was in a dye-sensitized cell using $CsSnI_3$ as a p-type hole transport layer and absorber. A common concern is the inclusion of lead as a component of the perovskite materials; solar cells based on tin-based perovskite absorbers such as $CH_3NH_3SnI_3$ have also been reported with lower power-conversion efficiencies.

In another recent development, solar cells based on transition metal oxide perovskites and heterostructures thereof such as $LaVO_3/SrTiO_3$ are studied.

Rice University scientists have discovered a novel phenomenon of light-induced lattice expansion in perovskite materials.

In order to overcome the instability issues with lead-based organic perovskite materials in ambient air and reduce the use of lead, perovskite derivatives, such as Cs_2SnI_6 double perovskite, have also been investigated.

Processing

Perovskite solar cells hold an advantage over traditional silicon solar cells in the simplicity of their processing and their tolerance to internal defects. Traditional silicon cells require expensive, multistep processes, conducted at high temperatures (>1000 °C) in a high vacuum in special clean room facilities. Meanwhile, the organic-inorganic perovskite material can be manufactured with simpler wet chemistry techniques in a traditional lab environment. Most notably, methylammonium and formamidinium lead trihalides, also known as hybrid perovskites, have been created using a variety of solvent techniques, such as spin coating, slot-die coating, blade coating, spray coating, inkjet printing, and screen printing, electrodeposition, and vapor deposition techniques, all of which have the potential to be scaled up with relative feasibility except spin coating.

Deposition Methods

The solution-based processing method can be classified into one-step solution deposition and two-step solution deposition. In the one-step deposition, a perovskite precursor solution that is prepared by mixing lead halide and organic halide together, is directly deposited through various coating methods, such as spin, spray, blade, and slot-die coating, to form perovskite film. One-step deposition is simple, fast, and inexpensive but it's also more challenging to control the perovskite film uniformity and quality. In the two-step deposition, the lead halide film is first deposited then reacts with organic halide to form perovskite film. The reaction takes time to complete but it can be facilitated by adding Lewis-bases or partial organic halide into lead halide precursors. In two-step deposition method, the volume expansion during the conversion of lead halide to perovskite can fill any pinholes to realize a better film quality. The vapor phase deposition can also be categorized into physical vapor deposition (PVD) and chemical vapor deposition (CVD). PVD refers to the evaporation of a pervoskite or its precursor to form a thin perovskite film on the substrate, which is free of solvent. While CVD involves the reaction of organic halide vapor with the lead halide thin film to convert it into the perovskite film.

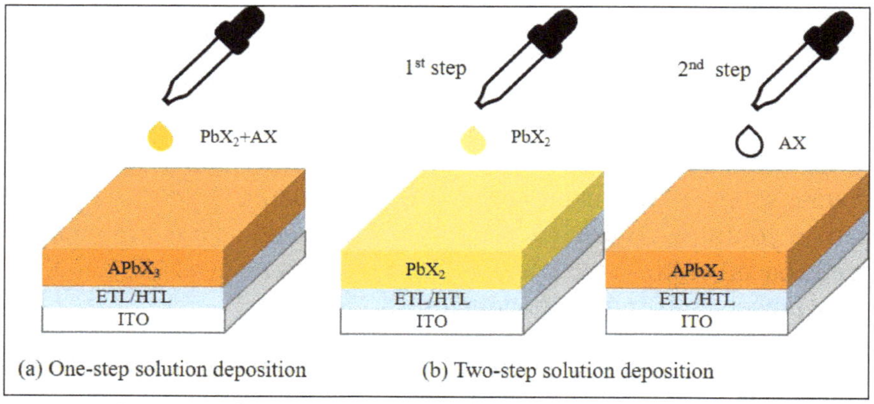

One-step solution deposition vs two-step solution deposition.

One-step Solution Deposition

In one-step solution processing, a lead halide and a methylammonium halide can be dissolved in a solvent and spin coated onto a substrate. Subsequent evaporation and convective self-assembly during spinning results in dense layers of well crystallized perovskite material, due to the strong ionic interactions within the material (The organic component also contributes to a lower crystallization temperature). However, simple spin-coating does not yield homogenous layers, instead requiring the addition of other chemicals such as GBL, DMSO, and toluene drips. Simple solution processing results in the presence of voids, platelets, and other defects in the layer, which would hinder the efficiency of a solar cell.

Another technique using room temperature solvent-solvent extraction produces high-quality crystalline films with precise control over thickness down to 20 nanometers across areas several centimeters square without generating pinholes. In this method "perovskite precursors are dissolved in a solvent called NMP and coated onto a substrate. Then, instead of heating, the substrate is bathed in diethyl ether, a second solvent that selectively grabs the NMP solvent and whisks it away. What's left is an ultra-smooth film of perovskite crystals."

In another solution processed method, the mixture of lead iodide and methylammonium halide dissolved in DMF is preheated. Then the mixture is spin coated on a substrate maintained at higher temperature. This method produces uniform films of up to 1 mm grain size.

Two-step Solution Deposition

In 2015, a new approach for forming the PbI_2 nanostructure and the use of high CH_3NH_3I concentration have been adopted to form high quality (large crystal size and smooth) perovskite film with better photovoltaic performances. On one hand, self-assembled porous PbI_2 is formed by incorporating small amounts of rationally chosen additives into the PbI_2 precursor solutions, which significantly facilitate the conversion of perovskite without any PbI_2 residue. On the other hand, through employing a relatively high CH_3NH_3I concentration, a firmly crystallized and uniform $CH_3NH_3PbI_3$ film is formed. Furthermore, this is an inexpensive approach.

Vapor Deposition

In vapor assisted techniques, spin coated or exfoliated lead halide is annealed in the presence of methylammonium iodide vapor at a temperature of around 150 °C. This technique holds an advantage over solution processing, as it opens up the possibility for multi-stacked thin films over larger areas. This could be applicable for the production of multi-junction cells. Additionally, vapor deposited techniques result in less thickness variation than simple solution processed layers. However, both techniques can result in planar thin film layers or for use in mesoscopic designs, such as coatings on a metal oxide scaffold. Such a design is common for current perovskite or dye-sensitized solar cells.

Scalability

Scalability includes not only scaling up the perovskite absorber layer, but also scaling up charge-transport layers and electrode. Both solution and vapor processes hold promise in terms of scalability. Process cost and complexity is significantly less than that of silicon solar cells. Vapor deposition or vapor assisted techniques reduce the need for use of further solvents, which reduces the risk of solvent remnants. Solution processing is cheaper. Current issues with perovskite solar cells revolve around stability, as the material is observed to degrade in standard environmental conditions, suffering drops in efficiency.

In 2014, Olga Malinkiewicz presented her inkjet printing manufacturing process for perovskite sheets in Boston (US) during the MRS fall meeting – for which she received MIT Technology review's innovators under 35 award. The University of Toronto also claims to have developed a low-cost Inkjet solar cell in which the perovskite raw materials are blended into a Nanosolar 'ink' which can be applied by an inkjet printer onto glass, plastic or other substrate materials.

Scaling Up the Absorber Layer

In order to scale up the perovskite layer while maintaining high efficiency, various techniques have been developed to coat the perovskite film more uniformly. For example, some physical approaches are developed to promote supersaturation through rapid solvent removal, thus getting more nucleations and reducing grain growth time and solute migration. Heating, gas flow, vacuum, and anti-solvent can all assist solvent removal. And chemical additives, such as chloride additives, Lewis base additives, surfacant additive, and surface modification, can influence the crystal growth to control the film mophology. For example, a recent report of surfacant additive, such as L-α-phosphatidylcholine (LP), demonstrated the suppression of solution flow by surfactants to eliminate gaps between islands and meanwhile the surface wetting improvement of perovskite ink on the hydrophobic substrate to ensure a full coverage. Besides, LP can also passivate charge traps to further enhance the device performance, which can be used in blade coating to get a high-throughput of PSCs with minimal efficiency loss.

Scaling Up the Charge-transport Layer

Scaling up the charge-transport layer is also necessary for the scalability of PSCs. Common electron transport layer (ETL) in n-i-p PSCs are TiO_2, SnO_2 and ZnO. Currently, to make TiO_2 layer deposition be compatiable with flexible polymer substrate, low-temperature techniques, such as atomic layer deposition, hydrothermal reaction, and electrodeposition, are developed to deposit compact TiO_2 layer in large area. Same methods also apply to SnO_2 deposition. As for hole transport layer (HTL), instead of commonly used PEDOT:PSS, NiO_x is used as an alternative due to the water absorption of PEDOT, which can be deposited through room-temperature solution processing. CuSCN is also an alternative HTL materials and can be deposited by spray coating, blade coating, and electrodeposition, which are potentially scalable. Researchers also report a molecular doping method for scalable blading to make HTL-free PSCs.

Scaling Up the Back Electrode

Evaporation deposition of back electrode is mature and scalable but it requires vacuum. Vacuum-free deposition of back electrode is important to fully imploit the solution processibility of PSCs. Silver electrodes can be screen-printed, and silver nanowire network can be spray-coated as back electrode. Carbon is also a potential candidate as scalable PSCs electrode, such as graphite, carbon nanotubes, and graphene.

Physics

An important characteristic of the most commonly used perovskite system, the methylammonium lead halides, is a bandgap controllable by the halide content. The materials also display a diffusion length for both holes and electrons of over one micron. The long diffusion length means that these materials can function effectively in a thin-film architecture, and that charges can be transported in the perovskite itself over long distances. It has recently been reported that charges in the perovskite material are predominantly present as free electrons and holes, rather than as bound excitons, since the exciton binding energy is low enough to enable charge separation at room temperature.

Efficiency Limits

Perovskite solar cell bandgaps are tunable and can be optimised for the solar spectrum by altering the halide content in the film (i.e., by mixing I and Br). The Shockley–Queisser limit radiative efficiency limit, also known as the detailed balance limit, is about 31% under an AM1.5G solar spectrum at 1000 W/m^2, for a Perovskite bandgap of 1.55 eV. This is slightly smaller than the radiative limit of gallium arsenide of bandgap 1.42 eV which can reach a radiative efficiency of 33%.

Values of the detailed balance limit are available in tabulated form and a MATLAB program for implementing the detailed balance model has been written. In the meantime, the drift-diffusion model has found to successfully predict the efficiency limit of perovskite solar cells, which enable us to understand the device physics in-depth, especially the radiative recombination limit and selective contact on device performance. There are two prerequisites for predicting and approaching the perovskite efficiency limit. First, the intrinsic radiative recombination needs to be corrected after adopting optical designs which will significantly affect the open-circuit voltage at its Shockley–Queisser limit. Second, the contact characteristics of the electrodes need to be carefully engineered to eliminate the charge accumulation and surface recombination at the electrodes. With the two procedures, the accurate prediction of efficiency limit and precise evaluation of efficiency degradation for perovskite solar cells are attainable by the drift-diffusion model.

Along with analytical calculations, there have been many first principle studies to find the characteristics of the perovskite material numerically. These include but are not limited to bandgap, effective mass, and defect levels for different perovskite materials. Also there have some efforts to cast light on the device mechanism based on simulations where Agrawal et al. suggests a modeling framework, presents analysis of near ideal efficiency, and talks about the importance of interface of perovskite and hole/electron transport layers. However, Sun et al. tries to come up with a compact model for perovskite different structures based on experimental transport data.

Architectures

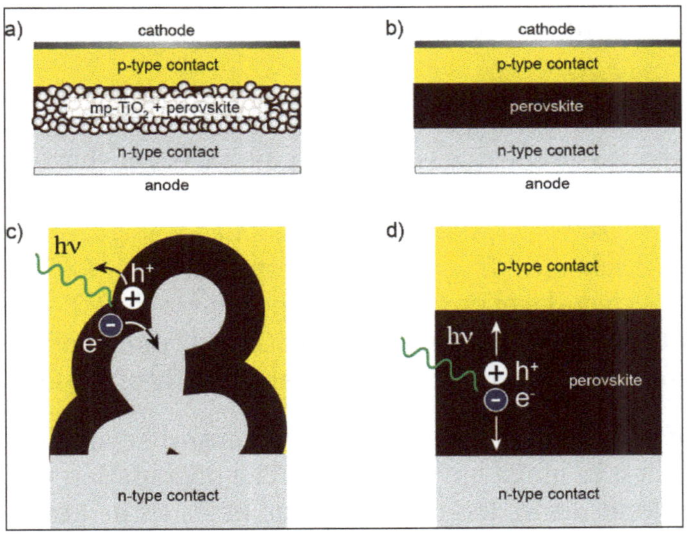

In figure, a) Schematic of a sensitized perovskite solar cell in which the active layer consist of a layer of mesoporous TiO_2 which is coated with the perovskite absorber. The active layer is contacted with an n-type material for electron extraction and a p-type material for hole extraction. b) Schematic of a thin-film perovskite solar cell. In this architecture in which just a flat layer of perovskite is sandwiched between two selective contacts. c) Charge generation and extraction in the sensitized architecture. After light absorption in the perovskite absorber the photogenerated electron is injected into the mesoporous TiO_2 through which it is extracted. The concomitantly generated hole is transferred to the p-type material. d) Charge generation and extraction in the thin-film architecture. After light absorption both charge generation as well as charge extraction occurs in the perovskite layer.

Perovskite solar cells function efficiently in a number of somewhat different architectures depending either on the role of the perovskite material in the device, or the nature of the top and bottom electrode. Devices in which positive charges are extracted by the transparent bottom electrode (cathode), can predominantly be divided into 'sensitized', where the perovskite functions mainly as a light absorber, and charge transport occurs in other materials, or 'thin-film', where most electron or hole transport occurs in the bulk of the perovskite itself. Similar to the sensitization in dye-sensitized solar cells, the perovskite material is coated onto a charge-conducting mesoporous scaffold – most commonly TiO_2 – as light-absorber. The photogenerated electrons are transferred from the perovskite layer to the mesoporous sensitized layer through which they are transported to the electrode and extracted into the circuit. The thin film solar cell architecture is based on the finding that perovskite materials can also act as highly efficient, ambipolar charge-conductor.

After light absorption and the subsequent charge-generation, both negative and positive charge carrier are transported through the perovskite to charge selective contacts. Perovskite solar cells emerged from the field of dye-sensitized solar cells, so the sensitized architecture was that initially used, but over time it has become apparent that they function well, if not ultimately better, in a thin-film architecture. More recently, some researchers also successfully demonstrated the possibility of fabricating flexible devices with perovskites, which makes it more promising for flexible energy demand. Certainly, the aspect of UV-induced degradation in the sensitized architecture may be detrimental for the important aspect of long-term stability.

There is another different class of architectures, in which the transparent electrode at the bottom acts as cathode by collecting the photogenerated p-type charge carriers.

Stability

One big challenge for perovskite solar cells (PSCs) is the aspect of short-term and long-term stability. The instability of PSCs is mainly related to environmental influence (moisture and oxygen), thermal influence (intrinsic stability), heating under applied voltage, photo influence (Ultraviolet light) (Visible light) and mechanical fragility. Several studies about PSCs stability have been performed and some elements have been proven to be important to the PSCs stability. However, there is no standard "operational" stability protocol for PSCs. But a method to quantify the intrinsic chemical stability of hybrid halide perovskites has been recently proposed.

The water-solubility of the organic constituent of the absorber material make devices highly prone to rapid degradation in moist environments. The degradation which is caused by moisture can be reduced by optimizing the constituent materials, the architecture of the cell, the interfaces and the environment conditions during the fabrication steps. Encapsulating the perovskite absorber with a composite of carbon nanotubes and an inert polymer matrix has been demonstrated to successfully prevent the immediate degradation of the material when exposed to moist ambient air at elevated temperatures. However, no long term studies and comprehensive encapsulation techniques have yet been demonstrated for perovskite solar cells. Besides moisture instability, it has also been shown that the embodiment of devices in which a mesoporous TiO_2 layer is sensitized with the perovskite absorber exhibits UV light induced instability. The cause for the observed decline in device performance of those solar cells is linked to the interaction between photogenerated holes inside the TiO_2 and oxygen radicals on the surface of TiO_2.

The measured ultra low thermal conductivity of 0.5 W/(Km) at room temperature in $CH_3NH_3PbI_3$ can prevent fast propagation of the light deposited heat, and keep the cell resistive on thermal stresses that can reduce its life time. The PbI_2 residue in perovskite film has been experimentally demonstrated to have a negative effect on the long-term stability of devices. The stabilization problem is claimed to be solved by replacing the organic transport layer with a metal oxide layer, allowing the cell to retain 90% capacity after 60 days. Besides, the two instabilities issues can be solved by using multifunctional fluorinated photopolymer coatings that confer luminescent and easy-cleaning features on the front side of the devices, while concurrently forming a strongly hydrophobic barrier toward environmental moisture on the back contact side. The front coating can prevent the UV light of the whole incident solar spectrum from negatively interacting with the PSC stack by converting it into visible light, and the back layer can prevent water from permeation within the solar cell stack. The resulting devices demonstrated excellent stability in terms of power conversion efficiencies during a 180-day aging test in the lab and a real outdoor condition test for more than 3 months.

In July 2015, major hurdles were that the largest perovskite solar cell was only the size of a fingernail and that they degraded quickly in moist environments. However, researchers from EPFL published in June 2017, a work successfully demonstrating large scale perovskite solar modules with no observed degradation over one year. Now, together with other organizations, the research team aims to develop a fully printable perovskite solar cell with 22% efficiency and with 90% of performance after ageing tests.

The intrinsic fragility of the perovskite material requires extrinsic reinforcement to shield this crucial layer from mechanical stresses. Insertion of mechanically reinforcing scaffolds directly into the active layers of perovskite solar cells resulted in the compound solar cell formed exhibiting a 30-fold increase in fracture resistance, repositioning the fracture properties of perovskite solar cells into the same domain as conventional c-Si, CIGS and CdTe solar cells.

Hysteretic Current-voltage Behavior

Another major challenge for perovskite solar cells is the observation that current-voltage scans yield ambiguous efficiency values. The power conversion efficiency of a solar cell is usually determined by characterizing its current-voltage (IV) behavior under simulated solar illumination. In contrast to other solar cells, however, it has been observed that the IV-curves of perovskite solar cells show a hysteretic behavior: depending on scanning conditions – such as scan direction, scan speed, light soaking, biasing – there is a discrepancy between the scan from forward-bias to short-circuit (FB-SC) and the scan from short-circuit to forward bias (SC-FB). Various causes have been proposed such as ion movement, polarization, ferroelectric effects, filling of trap states, however, the exact origin for the hysteretic behavior is yet to be determined. But it appears that determining the solar cell efficiency from IV-curves risks producing inflated values if the scanning parameters exceed the time-scale which the perovskite system requires in order to reach an electronic steady-state. Two possible solutions have been proposed: Unger et al. show that extremely slow voltage-scans allow the system to settle into steady-state conditions at every measurement point which thus eliminates any discrepancy between the FB-SC and the SC-FB scan.

Henry Snaith et al. have proposed 'stabilized power output' as a metric for the efficiency of a solar cell. This value is determined by holding the tested device at a constant voltage around the maximum power-point (where the product of voltage and photocurrent reaches its maximum value) and track the power-output until it reaches a constant value. Both methods have been demon-strated to yield lower efficiency values when compared to efficiencies determined by fast IV-scans. However, initial studies have been published that show that surface passivation of the perovskite absorber is an avenue with which efficiency values can be stabilized very close to fast-scan efficiencies. No obvious hysteresis of photocurrent was observed by changing the sweep rates or the direction in devices or the sweep rates. This indicates that the origin of hysteresis in photocurrent is more likely due to the trap formation in some non optimized films and device fabrication processes. The ultimate way to examine the efficiency of a solar cell device is to measure its power output at the load point. If there is large density of traps in the devices or photocurrent hysteresis for other reasons, the photocurrent would rise slowly upon turning on illumination This suggests that the interfaces might play a crucial role with regards to the hysteretic IV behavior since the major difference of the inverted architecture to the regular architectures is that an organic n-type contact is used instead of a metal oxide.

The observation of hysteretic current-voltage characteristics has thus far been largely underreported. Only a small fraction of publications acknowledge the hysteretic behavior of the described devices, even fewer articles show slow non-hysteretic IV curves or stabilized power outputs. Reported efficiencies, based on rapid IV-scans, have to be considered fairly unreliable and make it currently difficult to genuinely assess the progress of the field.

The ambiguity in determining the solar cell efficiency from current-voltage characteristics due to the observed hysteresis has also affected the certification process done by accredited laboratories such as NREL. The record efficiency of 20.1% for perovskite solar cells accepted as certified value by NREL in November 2014, has been classified as 'not stabilized'. To be able to compare results from different institution, it is necessary to agree on a reliable measurement protocol, as it has been proposed by including the corresponding Matlab code which can be found at GitHub.

Perovskites for Tandem Applications

A perovskite cell combined with bottom cell such as Si or copper indium gallium selenide (CIGS) as a tandem design can suppress individual cell bottlenecks and take advantage of the complementary characteristics to enhance the efficiency. This type of cells have higher efficiency potential, and therefore attracted recently a large attention from academic researchers.

4-Terminal Tandems

Using a four terminal configuration in which the two sub-cells are electrically isolated, Bailie et al. obtained a 17% and 18.6% efficient tandem cell with mc-Si ($\eta \sim 11\%$) and copper indium gallium selenide (CIGS, $\eta \sim 17\%$) bottom cells, respectively. A 13.4% efficient tandem cell with a highly efficient a-Si:H/c-Si heterojunction bottom cell using the same configuration was obtained. The application of TCO-based transparent electrodes to perovskite cells allowed to fabricate near-infrared transparent devices with improved efficiency and lower parasitic absorption losses. The application of these cells in 4-terminal tandems allowed improved efficiencies up to 26.7% when using a silicon bottom cell and up to 23.9% with a CIGS bottom cell.

2-Terminal Tandems

Mailoa et al. started the efficiency race for monolithic 2-terminal tandems using an homojunction c-Si bottom cell and demonstrate a 13.7% cell, largely limited by parasitic absorption losses. Then, Albrecht et al. developed a low-temperature processed perovskite cells using a SnO_2 electron transport layer. This allowed the use of silicon heterojunction solar cells as bottom cell and tandem efficiencies up to 18.1%. Werner et al. then improved this performance replacing the SnO_2 layer with PCBM and introducing a sequential hybrid deposition method for the perovskite absorber, leading to a tandem cell with 21.2% efficiency. Important parasitic absorption losses due to the use of Spiro-OMeTAD were still limiting the overall performance. An important change was demonstrated by Bush et al., who inverted the polarity of the top cell (n-i-p to p-i-n). They used a bilayer of SnO_2 and zinc tin oxide (ZTO) processed by ALD to work as a sputtering buffer layer, which enables the following deposition of a transparent top indium tin oxide (ITO) electrode. This change helped to improve the environmental and thermal stability of the perovskite cell and was crucial to further improve the perovskite/silicon tandem performance to 23.6%.

In the continuity, using a p-i-n perovskite top cell, Sahli et al. demonstrated in June 2018 a fully textured monolithic tandem cell with 25.2% efficiency, independently certified by Fraunhofer ISE CalLab. This improved efficiency can largely be attributed to the massively reduced reflection losses (below 2% in the range 360 nm-1000 nm, excluding metallization) and reduced parasitic absorption losses, leading to certified short-circuit currents of 19.5 mA/cm^2. Also in June 2018 the company Oxford Photovoltaics presented a cell with 27.3% efficiency. The record currently stands at 28% as of December 2018.

Theoretical Modeling

There have been some efforts to predict the theoretical limits for these traditional tandem designs using a perovskite cell as top cell on a c-Si or a-Si/c-Si heterojunction bottom cell. To show that the output power can be even further enhanced, bifacial structures were studied as well. It was concluded that extra output power can be extracted from the bifacial structure as compared to a bifacial HIT cell when the albedo reflection takes on values between 10 and 40%, which are realistic.

Up-scaling

In May 2016, IMEC and its partner Solliance announced a tandem structure with a semi-transparent perovskite cell stacked on top of a back-contacted silicon cell. A combined power conversion efficiency of 20.2% was claimed, with the potential to exceed 30%.

All-perovskite Tandems

In 2016, the development of efficient low-bandgap (1.2 - 1.3 eV) perovskite materials and the fabrication of efficient devices based on these enabled a new concept: all-perovskite tandem solar cells, where two perovskite compounds with different bandgaps are stacked on top of each other. The first two- and four-terminal devices with this architecture reported in the literature achieved efficiencies of 17% and 20.3%. All-perovskite tandem cells offer the prospect of being the first fully solution-processable architecture that has a clear route to exceeding not only the efficiencies of silicon, but also GaAs and other expensive III-V semiconductor solar cells.

In 2017, Dewei Zhao et al. fabricated low-bandgap (~1.25 eV) mixed Sn-Pb perovskite solar cells (PVSCs) with the thickness of 620 nm, which enables larger grains and higher crystallinity to extend the carrier lifetimes to more than 250 ns, reaching a maximum power conversion efficiency (PCE) of 17.6%. Furthermore, this low-bandgap PVSC reached an external quantum efficiency (EQE) of more than 70% in the wavelength range of 700–900 nm, the essential infrared spectral region where sunlight transmitted to bottom cell. They also combined the bottom cell with a ~1.58 eV bandgap perovskite top cell to create an all-perovskite tandem solar cell with four terminals, obtaining a steady-state PCE of 21.0%, suggesting the possibility of fabricating high-efficiency all-perovskite tandem solar cells.

MULTI-JUNCTION SOLAR CELL

Multi-junction (MJ) solar cells are solar cells with multiple p–n junctions made of different semiconductor materials. Each material's p-n junction will produce electric current in response to different wavelengths of light. The use of multiple semiconducting materials allows the absorbance of a broader range of wavelengths, improving the cell's sunlight to electrical energy conversion efficiency.

Traditional single-junction cells have a maximum theoretical efficiency of 33.16%. Theoretically, an infinite number of junctions would have a limiting efficiency of 86.8% under highly concentrated sunlight.

Currently, the best lab examples of traditional crystalline silicon (c-Si) solar cells have efficiencies between 20% and 25%, while lab examples of multi-junction cells have demonstrated performance over 46% under concentrated sunlight. Commercial examples of tandem cells are widely available at 30% under one-sun illumination, and improve to around 40% under concentrated sunlight. However, this efficiency is gained at the cost of increased complexity and manufacturing price. To date, their higher price and higher price-to-performance ratio have limited their use to special roles, notably in aerospace where their high power-to-weight ratio is desirable. In terrestrial applications, these solar cells are emerging in concentrator photovoltaics (CPV), with a growing number of installations around the world.

Tandem fabrication techniques have been used to improve the performance of existing designs. In particular, the technique can be applied to lower cost thin-film solar cells using amorphous silicon, as opposed to conventional crystalline silicon, to produce a cell with about 10% efficiency that is lightweight and flexible. This approach has been used by several commercial vendors, but these products are currently limited to certain niche roles, like roofing materials.

Basics of Solar Cells

In figure, Band diagram illustration of the photovoltaic effect. Photons give their energy to electrons in the depletion or quasi-neutral regions. These move from the valence band to the conduction band. Depending on the location, electrons and holes are accelerated by E_{drift}, which gives generation photocurrent, or by E_{scatt}, which gives scattering photocurrent.

Traditional photovoltaic cells are commonly composed of doped silicon with metallic contacts deposited on the top and bottom. The doping is normally applied to a thin layer on the top of the cell, producing a p-n junction with a particular bandgap energy, E_g.

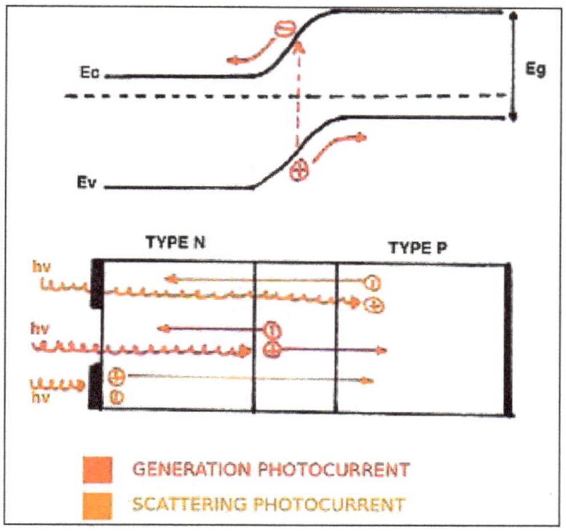

Photons that hit the top of the solar cell are either reflected or transmitted into the cell. Transmitted photons have the potential to give their energy, hν, to an electron if $h\nu \geq E_g$, generating an electron-hole pair. In the depletion region, the drift electric field E_{drift} accelerates both electrons and holes towards their respective n-doped and p-doped regions (up and down, respectively). The resulting current I_g is called the generated photocurrent. In the quasi-neutral region, the scattering electric field E_{scatt} accelerates holes (electrons) towards the p-doped (n-doped) region, which gives a scattering photocurrent I_{pscatt} (I_{nscatt}). Consequently, due to the accumulation of charges, a potential V and a photocurrent I_{ph} appear. The expression for this photocurrent is obtained by adding generation and scattering photocurrents: $I_{ph} = I_g + I_{nscatt} + I_{pscatt}$.

The J-V characteristics (J is current density, i.e. current per unit area) of a solar cell under illumination are obtained by shifting the J-V characteristics of a diode in the dark downward by I_{ph}. Since solar cells are designed to supply power and not absorb it, the power $P = V{\cdot}I_{ph}$ must be negative. Hence, the operating point (V_m, J_m) is located in the region where V>0 and I_{ph}<0, and chosen to maximize the absolute value of the power |P|.

Loss Mechanisms

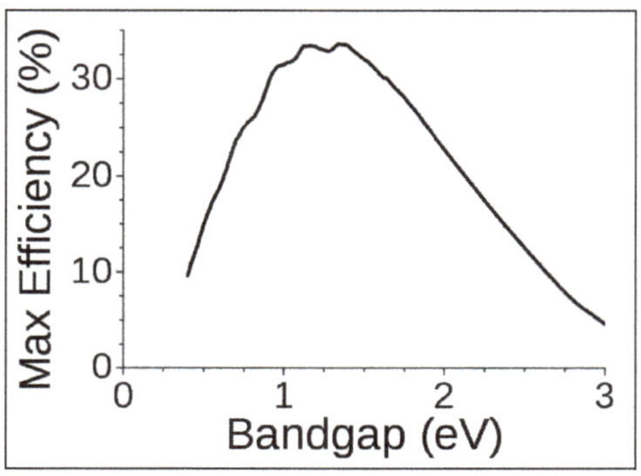

The Shockley-Queisser limit for the efficiency of a single-junction solar cell. It is essentially impossible for a single-junction solar cell, under unconcentrated sunlight, to have more than ~34% efficiency. A multi-junction cell, however, can exceed that limit.

The theoretical performance of a solar cell was first studied in depth in the 1960s, and is today known as the Shockley–Queisser limit. The limit describes several loss mechanisms that are inherent to any solar cell design.

The first are the losses due to blackbody radiation, a loss mechanism that affects any material object above absolute zero. In the case of solar cells at standard temperature and pressure, this loss accounts for about 7% of the power. The second is an effect known as "recombination", where the electrons created by the photoelectric effect meet the electron holes left behind by previous excitations. In silicon, this accounts for another 10% of the power.

However, the dominant loss mechanism is the inability of a solar cell to extract all of the power in the light, and the associated problem that it cannot extract any power at all from certain photons. This is due to the fact that the photons must have enough energy to overcome the bandgap of the material.

If the photon has less energy than the bandgap, it is not collected at all. This is a major consideration for conventional solar cells, which are not sensitive to most of the infrared spectrum, although that represents almost half of the power coming from the sun. Conversely, photons with more energy than the bandgap, say blue light, initially eject an electron to a state high above the bandgap, but this extra energy is lost through collisions in a process known as "relaxation". This lost energy turns into heat in the cell, which has the side-effect of further increasing blackbody losses.

Combining all of these factors, the maximum efficiency for a single-bandgap material, like conventional silicon cells, is about 34%. That is, 66% of the energy in the sunlight hitting the cell will be lost. Practical concerns further reduce this, notably reflection off the front surface or the metal terminals, with modern high-quality cells at about 22%.

Lower, also called narrower, bandgap materials will convert longer wavelength, lower energy photons. Higher, or wider bandgap materials will convert shorter wavelength, higher energy light. An analysis of the AM1.5 spectrum, shows the best balance is reached at about 1.1 eV (about 1100 nm, in the near infrared), which happens to be very close to the natural bandgap in silicon and a number of other useful semiconductors.

Multi-junction Cells

Cells made from multiple materials layers can have multiple bandgaps and will therefore respond to multiple light wavelengths, capturing and converting some of the energy that would otherwise be lost to relaxation.

For instance, if one had a cell with two bandgaps in it, one tuned to red light and the other to green, then the extra energy in green, cyan and blue light would be lost only to the bandgap of the green-sensitive material, while the energy of the red, yellow and orange would be lost only to the bandgap of the red-sensitive material. Following analysis similar to those performed for single-bandgap devices, it can be demonstrated that the perfect bandgaps for a two-gap device are at 1.1 eV and 1.8 eV.

Conveniently, light of a particular wavelength does not interact strongly with materials that are of bigger bandgap. This means that you can make a multi-junction cell by layering the different materials on top of each other, shortest wavelengths (biggest bandgap) on the "top" and increasing through the body of the cell. As the photons have to pass through the cell to reach the proper layer to be absorbed, transparent conductors need to be used to collect the electrons being generated at each layer.

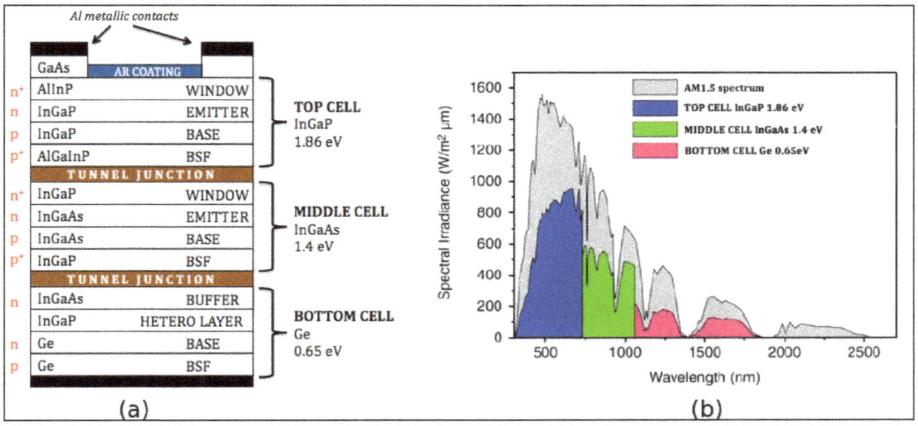

In figure, (a) The structure of an MJ solar cell. There are six important types of layers: pn junctions, back surface field (BSF) layers, window layers, tunnel junctions, anti-reflective coating and metallic contacts. (b) Graph of spectral irradiance E vs. wavelength λ over the AM1.5 solar spectrum, together with the maximum electricity conversion efficiency for every junction as a function of the wavelength.

Producing a tandem cell is not an easy task, largely due to the thinness of the materials and the difficulties extracting the current between the layers. The easy solution is to use two mechanically separate thin film solar cells and then wire them together separately outside the cell. This technique is widely used by amorphous silicon solar cells, Uni-Solar's products use three such layers to reach efficiencies around 9%. Lab examples using more exotic thin-film materials have demonstrated efficiencies over 30%.

The more difficult solution is the "monolithically integrated" cell, where the cell consists of a number of layers that are mechanically and electrically connected. These cells are much more difficult to produce because the electrical characteristics of each layer have to be carefully matched. In particular, the photocurrent generated in each layer needs to be matched, otherwise electrons will be absorbed between layers. This limits their construction to certain materials, best met by the III-V semiconductors.

Material Choice

The choice of materials for each sub-cell is determined by the requirements for lattice-matching, current-matching, and high performance opto-electronic properties.

For optimal growth and resulting crystal quality, the crystal lattice constant a of each material must be closely matched, resulting in lattice-matched devices. This constraint has been relaxed somewhat in recently developed metamorphic solar cells which contain a small degree of lattice mismatch. However, a greater degree of mismatch or other growth imperfections can lead to crystal defects causing a degradation in electronic properties.

Since each sub-cell is connected electrically in series, the same current flows through each junc-tion. The materials are ordered with decreasing bandgaps, E_g, allowing sub-bandgap light ($hc/\lambda < e\cdot Eg$)

to transmit to the lower sub-cells. Therefore, suitable bandgaps must be chosen such that the design spectrum will balance the current generation in each of the sub-cells, achieving current matching. Figure C(b) plots spectral irradiance E(λ), which is the source power density at a given wavelength λ. It is plotted together with the maximum conversion efficiency for every junction as a function of the wavelength, which is directly related to the number of photons available for conversion into photocurrent.

Finally, the layers must be electrically optimal for high performance. This necessitates usage of materials with strong absorption coefficients α(λ), high minority carrier lifetimes $\tau_{minority}$, and high mobilities μ.

The favorable values in the table below justify the choice of materials typically used for multi-junction solar cells: InGaP for the top sub-cell (E_g = 1.8 – 1.9 eV), InGaAs for the middle sub-cell (E_g = 1.4 eV), and Germanium for the bottom sub-cell (E_g = 0.67 eV). The use of Ge is mainly due to its lattice constant, robustness, low cost, abundance, and ease of production.

Because the different layers are closely lattice-matched, the fabrication of the device typically employs metal-organic chemical vapor deposition (MOCVD). This technique is preferable to the molecular beam epitaxy (MBE) because it ensures high crystal quality and large scale production.

Material	E_g, eV	a, nm	Absorption (λ = 0.8 μm), 1/μm	μ_n, $cm^2/(V{\cdot}s)$	τ_p, μs	Hardness (Mohs)	α, μm/K	S, m/s
c-Si	1.12	0.5431	0.102	1400	1	7	2.6	0.1–60
InGaP	1.86	0.5451	2	500	–	5	5.3	50
GaAs	1.4	0.5653	0.9	8500	3	4–5	6	50
Ge	0.65	0.5657	3	3900	1000	6	7	1000
InGaAs	1.2	0.5868	30	1200	–	–	5.66	100–1000

Structural Elements

Metallic Contacts

The metallic contacts are low-resistivity electrodes that make contact with the semiconductor layers. They are often aluminum. This provides an electrical connection to a load or other parts of a solar cell array. They are usually on two sides of the cell. And are important to be on the back face so that shadowing on the lighting surface is reduced.

Anti-reflective Coating

Anti-reflective (AR) coating is generally composed of several layers in the case of MJ solar cells. The top AR layer has usually a NaOH surface texturation with several pyramids in order to increase the transmission coefficient T, the trapping of the light in the material (because photons cannot easily get out the MJ structure due to pyramids) and therefore, the path length of photons in the material. On the one hand, the thickness of each AR layer is chosen to get destructive interferences. Therefore, the reflection coefficient R decreases to 1%. In the case of two AR layers L_1 (the top layer, usually SiO_2) and L_2 (usually TiO_2), there must be $n_{L_2} = n_{AlInP}^{1/2} \cdot n_{L_1}$ to have the same amplitudes for reflected fields and nL1dL1 = 4λmin,$nL_2dL_2 = \lambda_{min}/4$ to have opposite phase for reflected fields. On the other hand, the thickness of each AR layer is also chosen to minimize the reflectance at wavelengths for which the photocurrent is the lowest. Consequently, this maximizes JSC by matching currents of the three subcells. As example, because the current generated by the bottom cell is greater than the currents generated by the other cells, the thickness of AR layers is adjusted so that the infrared (IR) transmission (which corresponds to the bottom cell) is degraded while the ultraviolet transmission (which corresponds to the top cell) is upgraded. Particularly, an AR coating is very important at low wavelengths because, without it, T would be strongly reduced to 70%.

Tunnel Junctions

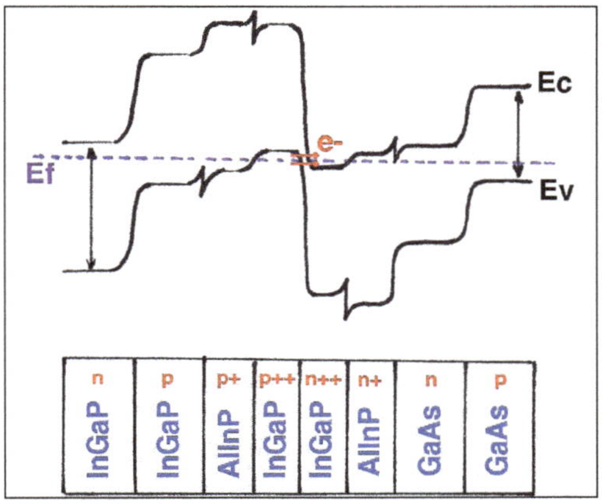

Layers and band diagram of the tunnel junction. Because the length of the depletion region is narrow and the band gap is high, electrons can tunnel.

The main goal of tunnel junctions is to provide a low electrical resistance and optically low-loss connection between two subcells. Without it, the p-doped region of the top cell would be directly connected with the n-doped region of the middle cell. Hence, a pn junction with opposite direction to the others would appear between the top cell and the middle cell. Consequently, the photovoltage would be lower than if there would be no parasitic diode. In order to decrease this effect, a tunnel junction is used. It is simply a wide band gap, highly doped diode. The high doping reduces the length of the depletion region because:

$$l_{depl} = \sqrt{\frac{2\epsilon(\phi_0 - V)}{q}\frac{N_A + N_D}{N_A N_D}}$$

Hence, electrons can easily tunnel through the depletion region. The J-V characteristic of the tunnel junction is very important because it explains why tunnel junctions can be used to have a low electrical resistance connection between two pn junctions. Figure D shows three different regions: the tunneling region, the negative differential resistance region and the thermal diffusion region. The region where electrons can tunnel through the barrier is called the tunneling region. There, the voltage must be low enough so that energy of some electrons who are tunneling is equal to energy states available on the other side of the barrier. Consequently, current density through the tunnel junction is high (with maximum value of J_P, the peak current density) and the slope near the origin is therefore steep. Then, the resistance is extremely low and consequently, the voltage too. This is why tunnel junctions are ideal for connecting two pn junctions without having a voltage drop. When voltage is higher, electrons cannot cross the barrier because energy states are no longer available for electrons. Therefore, the current density decreases and the differential resistance is negative. The last region, called thermal diffusion region, corresponds to the J-V characteristic of the usual diode:

$$J = J_S \left(\exp\left(\frac{qV}{kT} \right) - 1 \right)$$

In order to avoid the reduction of the MJ solar cell performances, tunnel junctions must be transparent to wavelengths absorbed by the next photovoltaic cell, the middle cell, i.e. $E_{gTunnel} > E_{gMiddleCell}$.

Window Layer and Back-surface Field

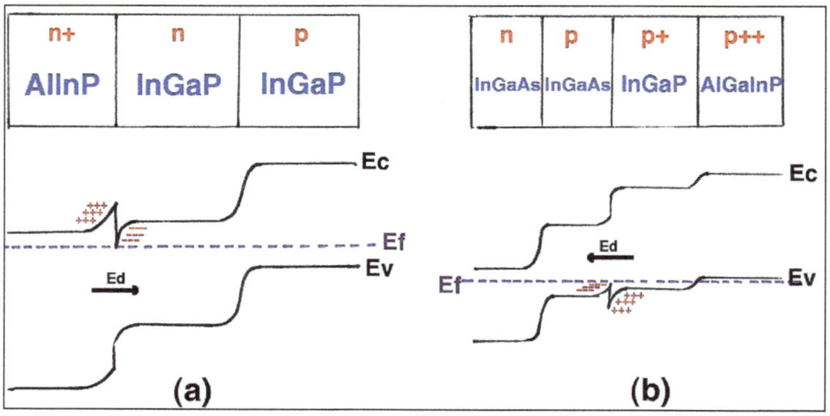

Layers and band diagram of a window layer. The surface recombination is reduced.
(b) Layers and band diagram of a BSF layer. The scattering of carriers is reduced.

A window layer is used in order to reduce the surface recombination velocity S. Similarly, a back-surface field (BSF) layer reduces the scattering of carriers towards the tunnel junction. The structure of these two layers is the same: it is a heterojunction which catches electrons (holes). Indeed, despite the electric field E_d, these cannot jump above the barrier formed by the heterojunction because they don't have enough energy, as illustrated in figure E. Hence, electrons (holes) cannot recombine with holes (electrons) and cannot diffuse through the barrier. By the way, window and BSF layers must be transparent to wavelengths absorbed by the next pn junction i.e. $E_{gWindow} > E_{gEmitter}$ and $E_{gBSF} > E_{gEmitter}$. Furthermore, the lattice constant must be close to the one of InGaP and the layer must be highly doped ($n \geq 10^{18}$ cm^{-3}).

J-V Characteristic

For maximum efficiency, each subcell should be operated at its optimal J-V parameters, which are not necessarily equal for each subcell. If they are different, the total current through the solar cell is the lowest of the three. By approximation, it results in the same relationship for the short-circuit current of the MJ solar cell: $J_{SC} = \min(J_{SC1}, J_{SC2}, J_{SC3})$ where $J_{SCi}(\lambda)$ is the short-circuit current density at a given wavelength λ for the subcell i.

Because of the impossibility to obtain J_{SC1}, J_{SC2}, J_{SC3} directly from the total J-V characteristic, the quantum efficiency QE(λ) is utilized. It measures the ratio between the amount of electron-hole pairs created and the incident photons at a given wavelength λ. Let $\varphi_i(\lambda)$ be the photon flux of corresponding incident light in subcell iand$QE_i(\lambda)$ be the quantum efficiency of the subcell i. By definition, this equates to:

$$QE_i(\lambda) = \frac{J_{SCi}(\lambda)}{q\phi_i(\lambda)} \Rightarrow J_{SCi} = \int_0^{\lambda 2} q\phi_i(\lambda) QE_i(\lambda) d\lambda$$

The value of is $QE_i(\lambda)$ obtained by linking it with the absorption coefficient $\alpha(\lambda)$, i.e. the number of photons absorbed per unit of length by a material. If it is assumed that each photon absorbed by a subcell creates an electron/hole pair (which is a good approximation), this leads to:

$$QE_i(\lambda) = 1 - e^{-\alpha(\lambda) d_i}$$

where d_i is the thickness of the subcell i and $e^{-\alpha(\lambda) d_i}$ is the percentage of incident light which is not absorbed by the subcell i.

Similarly, because,
$V = \sum_{i=1}^{3} V_i$, the following approximation can be used: $V_{OC} = \sum_{i=1}^{3} V_{OCi}$.
The values of V_{OCi} are then given by the J-V diode equation:

$$J_i = J_{0i}\left(e^{\frac{qV_i}{kT}} - 1\right) - J_{SCi} \Rightarrow V_{OCi} \approx \frac{kT}{q}\ln\left(\frac{J_{SCi}}{J_{0i}}\right)$$

Theoretical Limiting Efficiency

We can estimate the limiting efficiency of ideal infinite multi-junction solar cells using the graphical quantum-efficiency (QE) analysis invented by C. H. Henry. To fully take advantage of Henry's method, the unit of the AM1.5 spectral irradiance should be converted to that of photon flux (i.e., number of photons/m²/s). To do that, it is necessary to carry out an intermediate unit conversion from the power of electromagnetic radiation incident per unit area per photon energy to the photon flux per photon energy (i.e., from [W/m²/eV] to [number of photons/m²/s/eV]). For this intermediate unit conversion, the following points have to be considered: A photon has a distinct energy which is defined as follows:

$$E_{ph} = h \cdot f = h \cdot (c/\lambda)$$

where E_{ph} is photon energy, h is Planck's constant ($h = 6.626 * 10^{-34}$ [J·s]), c is speed of light ($c = 2.998 * 10^{8}$ [m/s]), f is frequency [1/s], and λ is wavelength [nm].

Then the photon flux per photon energy, $dn_{ph}/dh\nu$, with respect to certain irradiance E [W/m²/eV] can be calculated as follows:

$$\frac{dn_{ph}}{dh\nu} = \frac{E}{E_{ph}} = E/\{h \cdot (c/\lambda)\} = E[W/(m^2 \cdot eV)] \cdot \lambda \cdot (10^{-9}\,[m])/(1.998 \cdot 10^{-25}\,[J \cdot s \cdot m/s])$$

$$= E \cdot \lambda \cdot 5.03 \cdot 10^{15}\,[(\#\text{ of photons})/(m^2 \cdot s \cdot eV)]$$

As a result of this intermediate unit conversion, the AM1.5 spectral irradiance is given in unit of the photon flux per photon energy, [number of photons/m²/s/eV], as shown in figure.

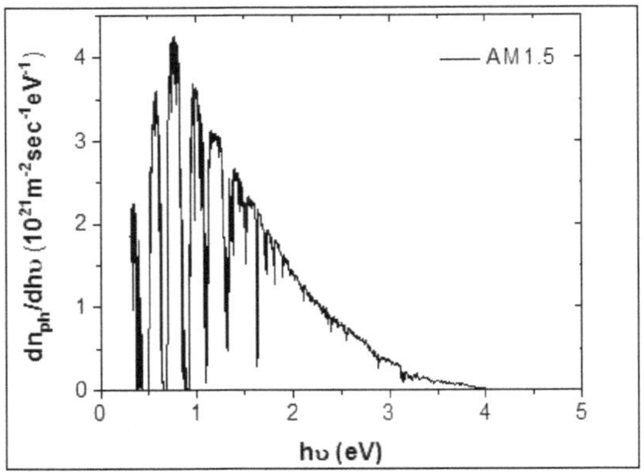

Photon flux per photon energy from standard solar energy spectrum (AM of 1.5).

Based on the above result from the intermediate unit conversion, we can derive the photon flux by numerically integrating the photon flux per photon energy with respect to photon energy. The numerically integrated photon flux is calculated using the Trapezoidal rule, as follows:

$$n_{ph}(E_g) = \int_{E_g}^{\infty} \frac{dn_{ph}}{dhv} dhv = \sum_{i=E_g}^{\infty} (hv_{i+1} - hv_i) \frac{1}{2} \left(\frac{dn_{ph}}{dhv}(hv_{i+1}) + \frac{dn_{ph}}{dhv}(hv_i) \right)$$

As a result of this numerical integration, the AM1.5 spectral irradiance is given in unit of the photon flux, [number of photons/m^2/s].

There are no photon flux data in the small photon energy range from 0 eV to 0.3096 eV because the standard (AM1.5) solar energy spectrum for hv < 0.31 eV are not available. Regardless of this data unavailability, however, the graphical QE analysis can be done using the only available data with a reasonable assumption that semiconductors are opaque for photon energies greater than their bandgap energy, but transparent for photon energies less than their bandgap energy. This assumption accounts for the first intrinsic loss in the efficiency of solar cells, which is caused by the inability of single-junction solar cells to properly match the broad solar energy spectrum. However, the current graphical QE analysis still cannot reflect the second intrinsic loss in the efficiency of solar cells, radiative recombination. To take the radiative recombination into account, we need to evaluate the radiative current density, J_{rad}, first. According to Shockley and Queisser method, J_{rad} can be approximated as follows:

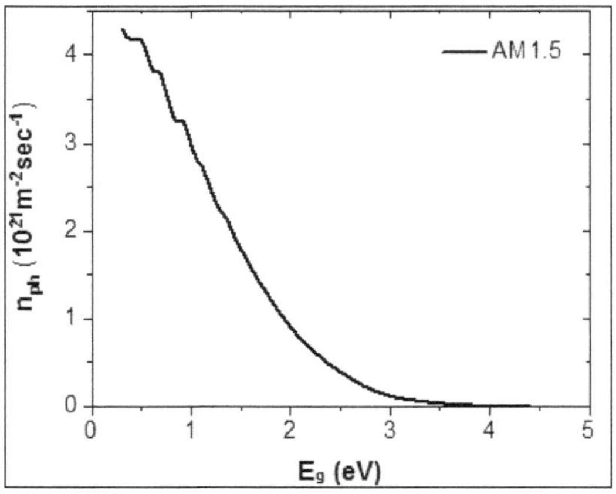

Photon flux from standard solar energy spectrum (AM of 1.5).

$$J_{rad} = A\,exp(\frac{eV - E_g}{kT})$$

$$A = \frac{2\pi exp(n^2 + 1)E_g^2 kT}{h^3 c^2}$$

where E_g is in electron volts and n is evaluated to be 3.6, the value for GaAs. The incident absorbed thermal radiation J_{th} is given by J_{rad} with V = 0.

$$J_{th} = A\,exp(\frac{-E_g}{kT})$$

The current density delivered to the load is the difference of the current densities due to absorbed solar and thermal radiation and the current density of radiation emitted from the top surface or absorbed in the substrate. Defining $J_{ph} = en_{ph}$, we have:

$$J = J_{ph} + J_{th} - J_{rad}$$

The second term, J_{th}, is negligible compared to J_{ph} for all semiconductors with E_g. ≥ 0.3 eV, as can be shown by evaluation of the above J_{th} equation. Thus, we will neglect this term to simplify the following discussion. Then we can express J as follows:

$$J = en_{ph} - Aexp(\frac{eV - E_g}{kT})$$

The open-circuit voltage is found by setting J = 0.

$$eV_{OC} = E_g - kTln(\frac{A}{en_{ph}})$$

The maximum power point (J_m, V_m) is found by stetting the derivative $\frac{dJV}{dV} = 0$. The familiar result of this calculation is:

$$eV_m = eV_{OC} - kTln(1 + \frac{eV_m}{kT})$$

$$J_m = \frac{en_{ph}}{1 + kT / eV_m}$$

Finally, the maximum work (W_m) done per absorbed photon, W_m is given by:

$$W_m = \frac{J_m V_m}{n_{ph}} = \frac{eV_m}{1 + kT / eV_m} = eV_m - kT$$

Combining the last three equations, we have:

$$W_m = E_g - kT[ln(\frac{A}{en_{ph}}) + ln(1 + \frac{eV_m}{kT}) + 1]$$

Using the above equation, W_m (red line) is plotted in figure for different values of E_g (or n_{ph}).

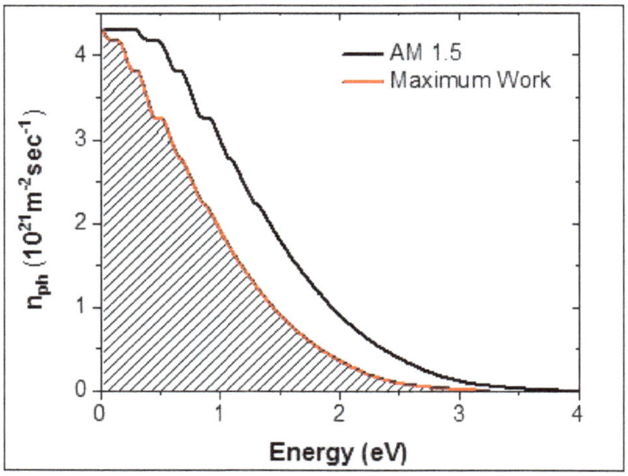

Maximum work by ideal infinite multi-junction solar cells under standard AM1.5 spectral irradiance.

Now, we can fully use Henry's graphical QE analysis, taking into account the two major intrinsic losses in the efficiency of solar cells. The two main intrinsic losses are radiative recombination, and the inability of single junction solar cells to properly match the broad solar energy spectrum. The shaded area under the red line represents the maximum work done by ideal infinite multi-junction solar cells. Hence, the limiting efficiency of ideal infinite multi-junction solar cells is evaluated to be 68.8% by comparing the shaded area defined by the red line with the total photon-flux area determined by the black line. (This is why this method is called "graphical" QE analysis.) Although this limiting efficiency value is consistent with the values published by Parrott and Vos in 1979: 64% and 68.2% respectively, there is a small gap between the estimated value in this report and literature values. This minor difference is most likely due to the different ways how to approximate the photon flux from 0 eV to 0.3096 eV. Here, we approximated the photon flux from 0 eV to 0.3096 eV as the same as the photon flux at 0.31 eV.

Materials

The majority of multi-junction cells that have been produced to date use three layers (although many tandem a-Si:H/mc-Si modules have been produced and are widely available). However, the triple junction cells require the use of semiconductors that can be tuned to specific frequencies, which has led to most of them being made of gallium arsenide (GaAs) compounds, often germanium for the bottom-, GaAs for the middle-, and $GaInP_2$ for the top-cell.

Gallium Arsenide Substrate

Dual junction cells can be made on Gallium arsenide wafers. Alloys of Indium gallium phosphide in the range $In_{.5}Ga_{.5}P$ through $In_{.53}Ga_{.47}P$ serve as the high band gap alloy. This alloy range provides for the ability to have band gaps in the range of 1.92eV to 1.87eV. The lower GaAs junction has a band gap of 1.42eV.

Germanium Substrate

Triple junction cells consisting of indium gallium phosphide (InGaP), gallium arsenide (GaAs) or indium gallium arsenide (InGaAs) and germanium (Ge) can be fabricated on germanium wafers. Early cells used straight gallium arsenide in the middle junction. Later cells have utilized $In_{0.015}Ga_{0.985}As$, due to the better lattice match to Ge, resulting in a lower defect density.

Due to the huge band gap difference between GaAs (1.42eV), and Ge (0.66eV), the current match is very poor, with the Ge junction operated significantly current limited.

Current efficiencies for commercial InGaP/GaAs/Ge cells approach 40% under concentrated sunlight. Lab cells (partly using additional junctions between the GaAs and Ge junction) have demonstrated efficiencies above 40%.

Indium Phosphide Substrate

Indium phosphide may be used as a substrate to fabricate cells with band gaps between 1.35eV and 0.74eV. Indium Phosphide has a band gap of 1.35eV. Indium gallium arsenide ($In_{0.53}Ga_{0.47}As$) is lattice matched to Indium Phosphide with a band gap of 0.74eV. A quaternary alloy of Indium gallium arsenide phosphide can be lattice matched for any band gap in between the two.

Indium phosphide-based cells have the potential to work in tandem with gallium arsenide cells. The two cells can be optically connected in series (with the InP cell below the GaAs cell), or in parallel through the use of spectra splitting using a Dichroic filter.

Indium Gallium Nitride Substrate

Indium gallium nitride (InGaN) is a semiconductor material made of a mix of gallium nitride (GaN) and indium nitride (InN). It is a ternary group III/V direct bandgap semiconductor. Its bandgap can be tuned by varying the amount of indium in the alloy from 0.7 eV to 3.4 eV, thus making it an ideal material for solar cells. However, its conversion efficiencies because of technological factors unrelated to bandgap are still not high enough to be competitive in the market.

Performance Improvements

Structure

Many MJ photovoltaic cells use III-V semiconductor materials. GaAsSb-based heterojunction tunnel diodes, instead of conventional InGaP highly doped tunnel diodes described above, have a lower tunneling distance. Indeed, in the heterostructure formed by GaAsSb and InGaAs, the valence band of GaAsSb is higher than the valence band of the adjoining p-doped layer. Consequently, the tunneling distance d_{tunnel} is reduced and so the tunneling current, which exponentially depends of d_{tunnel}, is increased. Hence, the voltage is lower than that of the InGaP tunnel junction. GaAsSb heterojunction tunnel diodes offer other advantages. The same current can be achieved by using a lower doping. Secondly, because the lattice constant is larger for GaAsSb than Ge, one can use a wider range of materials for the bottom cell because more materials are lattice-matched to GaAsSb than to Ge.

Chemical components can be added to some layers. Adding about one percent of Indium in each layer better matches lattice constants of the different layers. Without it, there is about 0.08 percent of mismatching between layers, which inhibits performance. Adding aluminium to the top cell increases its band gap to 1.96 eV, covering a larger part of the solar spectrum and obtain a higher open-circuit voltage V_{OC}.

The theoretical efficiency of MJ solar cells is 86.8% for an infinite number of pn junctions, implying that more junctions increase efficiency. The maximum theoretical efficiency is 37, 50, 56, 72% for 1, 2, 3, 36 pn junctions, respectively, with the number of junctions increasing exponentially to achieve equal efficiency increments. The exponential relationship implies that as the cell approaches the limit of efficiency, the increase cost and complexity grow rapidly. Decreasing the thickness of the top cell increases the transmission coefficient T.

Finally, an InGaP hetero-layer between the p-Ge layer and the InGaAs layer can be added in order to create automatically the n-Ge layer by scattering during MOCVD growth and increase significantly the quantum efficiency QE(λ) of the bottom cell. InGaP is advantageous because of its high scattering coefficient and low solubility in Ge.

Spectral Variations

Solar spectrum at the Earth surface changes constantly depending on the weather and sun position. This results in the variation of $\varphi(\lambda)$, $QE(\lambda)$, $\alpha(\lambda)$ and thus the short-circuit currents J_{SCi}. As a result, the current densities J_i are not necessarily matched and the total current becomes lower. These variations can be quantified using the average photon energy (APE) which is the ratio between the spectral irradiance G(λ) (the power density of the light source in a specific wavelength λ) and the total photon flux density. It can be shown that a high (low) value for APE means low (high) wavelengths spectral conditions and higher (lower) efficiencies. Thus APE is a good indicator for quantifying the effects of the solar spectrum variations on performances and has the added advantage of being independent of the device structure and the absorption profile of the device.

Use of Light Concentrators

Light concentrators increase efficiencies and reduce the cost/efficiency ratio. The three types of light concentrators in use are refractive lenses like Fresnel lenses, reflective dishes (parabolic or cassegraine), and light guide optics. Thanks to these devices, light arriving on a large surface can be concentrated on a smaller cell. The intensity concentration ratio (or "suns") is the average intensity of the focused light divided by 1 kW/m^2 (reasonable value related to solar constant). If its value is X then the MJ current becomes X higher under concentrated illumination.

Using concentrations on the order of 500 to 1000, meaning that a 1 cm^2 cell can use the light collected from 0.1 m^2 (as 1 m^2 equal 10000 cm^2), produces the highest efficiencies seen to date. Three-layer cells are fundamentally limited to 63%, but existing commercial prototypes have already demonstrated over 40%. These cells capture about 2/3 of their theoretical maximum performance, so assuming the same is true for a non-concentrated version of the same design, one might expect a three-layer cell of 30% efficiency. This is not enough of an advantage over traditional silicon designs to make up for their extra production costs. For this reason, almost all multi-junction cell research for terrestrial use is dedicated to concentrator systems, normally using mirrors or fresnel lenses.

Using a concentrator also has the added benefit that the number of cells needed to cover a given amount of ground area is greatly reduced. A conventional system covering 1 m^2 would require 625

16 cm^2 cells, but for a concentrator system only a single cell is needed, along with a concentrator. The argument for concentrated Multi-junction cells has been that the high cost of the cells themselves would be more than offset by the reduction in total number of cells. However, the downside of the concentrator approach is that efficiency drops off very quickly under lower lighting conditions. In order to maximize its advantage over traditional cells and thus be cost competitive, the concentrator system has to track the sun as it moves to keep the light focused on the cell and maintain maximum efficiency as long as possible. This requires a solar tracker system, which increases yield, but also cost.

Fabrication

As of 2014 multi-junction cells were expensive to produce, using techniques similar to semiconductor device fabrication, usually metalorganic vapour phase epitaxy but on "chip" sizes on the order of centimeters.

A new technique was announced that year that allowed such cells to use a substrate of glass or steel, lower-cost vapors in reduced quantities that was claimed to offer costs competitive with conventional silicon cells.

Comparison with other Technologies

There are four main categories of photovoltaic cells: conventional mono and multi crystalline silicon (c-Si) cells, thin film solar cells (a-Si, CIGS and CdTe), and multi-junction (MJ) solar cells. The fourth category, emerging photovoltaics, contains technologies that are still in the research or development phase and are not listed in the table below.

Categories	Technology	η (%)	V_{OC} (V)	I_{SC} (A)	W/m^2	t (μm)
Crystalline silicon cells	Monocrystalline	24.7	0.5	0.8	63	100
	Polysilicon	20.3	0.615	8.35	211	200
Thin film solar cells	Amorphous silicon	11.1	0.63	0.089	33	1
	CdTe	16.5	0.86	0.029	–	5
	CIGS	19.5	–	–	–	1
Multi-junction cells	MJ	40.7	2.6	1.81	476	140

MJ solar cells and other photovoltaic devices have significant differences. Physically, the main property of a MJ solar cell is having more than one pn junction in order to catch a larger photon energy spectrum while the main property of the thin film solar cell is to use thin films instead of thick layers in order to decrease the cost efficiency ratio. As of 2010, MJ solar panels are more expensive than others. These differences imply different applications: MJ solar cells are preferred in space and c-Si solar cells for terrestrial applications.

The efficiencies of solar cells and Si solar technology are relatively stable, while the efficiency of solar modules and multi-junction technology are progressing.

Measurements on MJ solar cells are usually made in laboratory, using light concentrators (this is often not the case for the other cells) and under standard test conditions (STCs). STCs prescribe, for terrestrial applications, the AM1.5 spectrum as the reference. This air mass (AM) corresponds to a fixed position of the sun in the sky of 48° and a fixed power of 833 W/m^2.

Therefore, spectral variations of incident light and environmental parameters are not taken into account under STC.

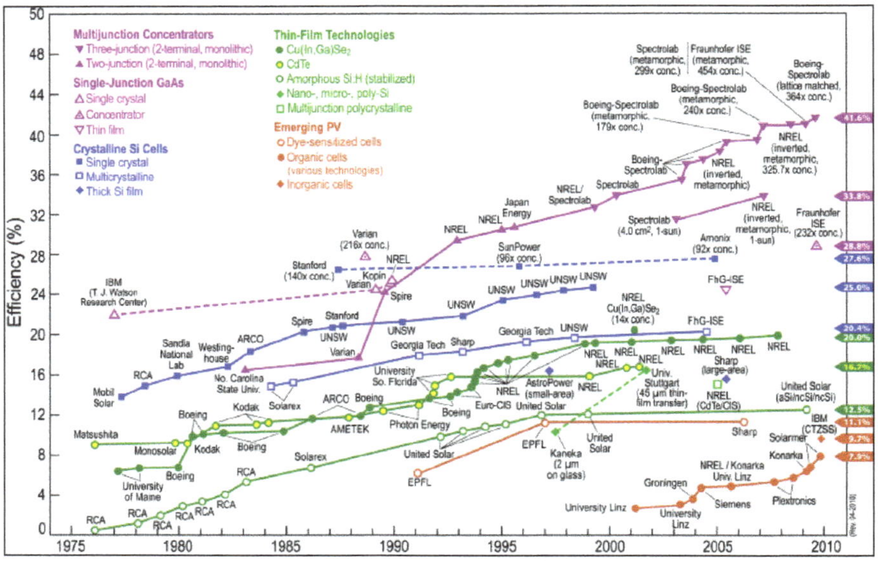

National Renewable Energy Laboratory graph of solar cell efficiency over time.

Consequently, performance of MJ solar cells in terrestrial environment is inferior to that achieved in laboratory. Moreover, MJ solar cells are designed such that currents are matched under STC, but not necessarily under field conditions. One can use QE(λ) to compare performances of different technologies, but QE(λ) contains no information on the matching of currents of subcells. An important comparison point is rather the output power per unit area generated with the same incident light.

Applications

As of 2010, the cost of MJ solar cells was too high to allow use outside of specialized applications. The high cost is mainly due to the complex structure and the high price of materials. Nevertheless, with light concentrators under illumination of at least 400 suns, MJ solar panels become practical.

As less expensive multi-junction materials become available other applications involve bandgap engineering for microclimates with varied atmospheric conditions.

MJ cells are currently being utilized in the Mars rover missions.

The environment in space is quite different. Because there is no atmosphere, the solar spectrum is different (AM0). The cells have a poor current match due to a greater photon flux of photons above 1.87eV vs. those between 1.87eV and 1.42eV. This results in too little current in the GaAs junction, and hampers the overall efficiency since the InGaP junction operates below MPP current and the GaAs junction operates above MPP current. To improve current match, the InGaP layer is intentionally thinned to allow additional photons to penetrate to the lower GaAs layer.

In terrestrial concentrating applications, the scatter of blue light by the atmosphere reduces the photon flux above 1.87eV, better balancing the junction currents. Radiation particles that are no longer filtered can damage the cell. There are two kinds of damage: ionisation and atomic displacement.

Still, MJ cells offer higher radiation resistance, higher efficiency and a lower temperature coefficient.

DYE-SENSITIZED SOLAR CELL

A dye-sensitized solar cell (DSSC, DSC, DYSC or Grätzel cell) is a low-cost solar cell belonging to the group of thin film solar cells. It is based on a semiconductor formed between a photo-sensitized anode and an electrolyte, a photoelectrochemical system. The modern version of a dye solar cell, also known as the Grätzel cell, was originally co-invented in 1988 by Brian O'Regan and Michael Grätzel at UC Berkeley and this work was later developed by the aforementioned scientists at the École Polytechnique Fédérale de Lausanne until the publication of the first high efficiency DSSC in 1991. Michael Grätzel has been awarded the 2010 Millennium Technology Prize for this invention.

A selection of dye-sensitized solar cells.

The DSSC has a number of attractive features; it is simple to make using conventional roll-printing techniques, is semi-flexible and semi-transparent which offers a variety of uses not applicable to glass-based systems, and most of the materials used are low-cost. In practice it has proven difficult to eliminate a number of expensive materials, notably platinum and ruthenium, and the liquid electrolyte presents a serious challenge to making a cell suitable for use in all weather. Although its conversion efficiency is less than the best thin-film cells, in theory its price/performance ratio should be good enough to allow them to compete with fossil fuel electrical generation by achieving grid parity. Commercial applications, which were held up due to chemical stability problems, are forecast in the European Union Photovoltaic Roadmap to significantly contribute to renewable electricity generation by 2020.

Current Technology: Semiconductor Solar Cells

In a traditional solid-state semiconductor, a solar cell is made from two doped crystals, one doped with n-type impurities (n-type semiconductor), which add additional free conduction band electrons, and the other doped with p-type impurities (p-type semiconductor), which add additional electron holes. When placed in contact, some of the electrons in the n-type portion flow into the p-type to "fill in" the missing electrons, also known as electron holes. Eventually enough electrons will flow across the boundary to equalize the Fermi levels of the two materials. The result is a region at the interface, the p-n junction, where charge carriers are depleted and/or accumulated on each side of the interface. In silicon, this transfer of electrons produces a potential barrier of about 0.6 to 0.7 V.

When placed in the sun, photons of the sunlight can excite electrons on the p-type side of the semiconductor, a process known as photoexcitation. In silicon, sunlight can provide enough energy to push an electron out of the lower-energy valence band into the higher-energy conduction band. As the name implies, electrons in the conduction band are free to move about the silicon. When a load is placed across the cell as a whole, these electrons will flow out of the p-type side into the n-type side, lose energy while moving through the external circuit, and then flow back into the p-type material where they can once again re-combine with the valence-band hole they left behind. In this way, sunlight creates an electric current.

In any semiconductor, the band gap means that only photons with that amount of energy, or more, will contribute to producing a current. In the case of silicon, the majority of visible light from red to violet has sufficient energy to make this happen. Unfortunately higher energy photons, those at the blue and violet end of the spectrum, have more than enough energy to cross the band gap; although some of this extra energy is transferred into the electrons, the majority of it is wasted as heat. Another issue is that in order to have a reasonable chance of capturing a photon, the n-type layer has to be fairly thick. This also increases the chance that a freshly ejected electron will meet up with a previously created hole in the material before reaching the p-n junction. These effects produce an upper limit on the efficiency of silicon solar cells, currently around 12 to 15% for common modules and up to 25% for the best laboratory cells (33.16% is the theoretical maximum efficiency for single band gap solar cells).

By far the biggest problem with the conventional approach is cost; solar cells require a relatively thick layer of doped silicon in order to have reasonable photon capture rates, and silicon processing is expensive. There have been a number of different approaches to reduce this cost over the last decade, notably the thin-film approaches, but to date they have seen limited application due to a variety of practical problems. Another line of research has been to dramatically improve efficiency through the multi-junction approach, although these cells are very high cost and suitable only for large commercial deployments. In general terms the types of cells suitable for rooftop deployment have not changed significantly in efficiency, although costs have dropped somewhat due to increased supply.

Dye-sensitized Solar Cells

In the late 1960s it was discovered that illuminated organic dyes can generate electricity at oxide electrodes in electrochemical cells. In an effort to understand and simulate the primary processes in photosynthesis the phenomenon was studied at the University of California at Berkeley with chlorophyll extracted from spinach (bio-mimetic or bionic approach). On the basis of such experiments electric power generation via the dye sensitization solar cell (DSSC) principle was demonstrated and discussed in 1972. The instability of the dye solar cell was identified as a main challenge. Its efficiency could, during the following two decades, be improved by optimizing the porosity of the electrode prepared from fine oxide powder, but the instability remained a problem.

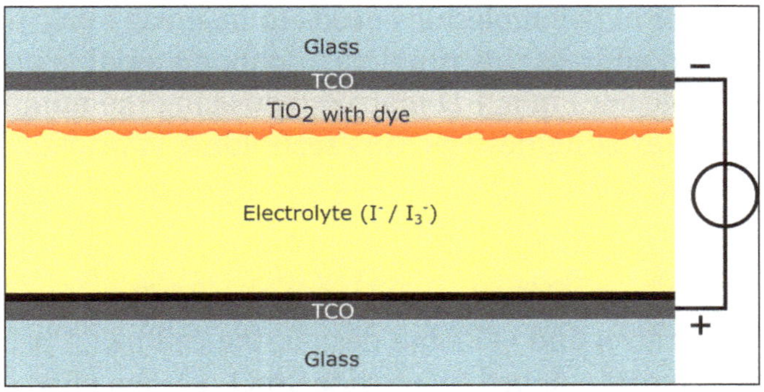

Type of cell made at the EPFL.

A modern DSSC is composed of a porous layer of titanium dioxide nanoparticles, covered with a molecular dye that absorbs sunlight, like the chlorophyll in green leaves. The titanium dioxide is immersed under an electrolyte solution, above which is a platinum-based catalyst. As in a conventional alkaline battery, an anode (the titanium dioxide) and a cathode (the platinum) are placed on either side of a liquid conductor (the electrolyte).

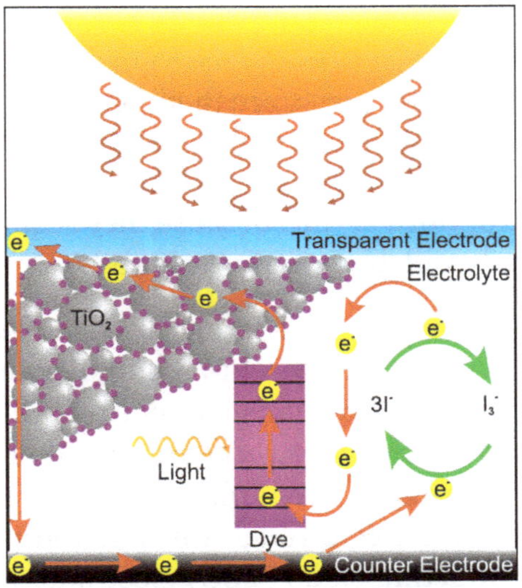

Operation of a Grätzel cell.

Sunlight passes through the transparent electrode into the dye layer where it can excite electrons that then flow into the titanium dioxide. The electrons flow toward the transparent electrode where they are collected for powering a load. After flowing through the external circuit, they are re-introduced into the cell on a metal electrode on the back, flowing into the electrolyte. The electrolyte then transports the electrons back to the dye molecules.

Dye-sensitized solar cells separate the two functions provided by silicon in a traditional cell design. Normally the silicon acts as both the source of photoelectrons, as well as providing the electric field to separate the charges and create a current. In the dye-sensitized solar cell, the bulk of the semiconductor is used solely for charge transport, the photoelectrons are provided from a separate photosensitive dye. Charge separation occurs at the surfaces between the dye, semiconductor and electrolyte.

The dye molecules are quite small (nanometer sized), so in order to capture a reasonable amount of the incoming light the layer of dye molecules needs to be made fairly thick, much thicker than the molecules themselves. To address this problem, a nanomaterial is used as a scaffold to hold large numbers of the dye molecules in a 3-D matrix, increasing the number of molecules for any given surface area of cell. In existing designs, this scaffolding is provided by the semiconductor material, which serves double-duty.

Construction

In the case of the original Grätzel and O'Regan design, the cell has 3 primary parts. On top is a transparent anode made of fluoride-doped tin dioxide (SnO_2:F) deposited on the back of a (typically glass) plate. On the back of this conductive plate is a thin layer of titanium dioxide (TiO_2), which forms into a highly porous structure with an extremely high surface area. The (TiO_2) is chemically bound by a process called sintering. TiO_2 only absorbs a small fraction of the solar photons (those in the UV). The plate is then immersed in a mixture of a photosensitive ruthenium-polypyridine dye (also called molecular sensitizers) and a solvent. After soaking the film in the dye solution, a thin layer of the dye is left covalently bonded to the surface of the TiO_2. The bond is either an ester, chelating, or bidentate bridging linkage.

A separate plate is then made with a thin layer of the iodide electrolyte spread over a conductive sheet, typically platinum metal. The two plates are then joined and sealed together to prevent the electrolyte from leaking. The construction is simple enough that there are hobby kits available to hand-construct them. Although they use a number of "advanced" materials, these are inexpensive compared to the silicon needed for normal cells because they require no expensive manufacturing steps. TiO_2, for instance, is already widely used as a paint base.

One of the efficient DSSCs devices uses ruthenium-based molecular dye, e.g. [Ru(4,4'-dicarboxy-2,2'-bipyridine)$_2$(NCS)$_2$] (N3), that is bound to a photoanode via carboxylate moieties. The photoanode consists of 12 μm thick film of transparent 10–20 nm diameter TiO_2 nanoparticles covered with a 4 μm thick film of much larger (400 nm diameter) particles that scatter photons back into the transparent film. The excited dye rapidly injects an electron into the TiO_2 after light absorption. The injected electron diffuses through the sintered particle network to be collected at the front side transparent conducting oxide (TCO) electrode, while the dye is regenerated via reduction by a redox shuttle, I_3^-/I^-, dissolved in a solution. Diffusion of the oxidized form of the shuttle to the counter electrode completes the circuit.

Mechanism of DSSCs

The following steps convert in a DSSC photons (light) to current:

- The incident photon is absorbed by Ru complex photosensitizers adsorbed on the TiO_2 surface.
- The photosensitizers are excited from the ground state (S) to the excited state (S*). The excited electrons are injected into the conduction band of the TiO_2 electrode. This results in the oxidation of the photosensitizer (S^+).

$$S + h\nu \rightarrow S^{\cdot}$$

$$S^{\cdot} \xrightarrow{TiO_2} S^+ + e^-$$

- The injected electrons in the conduction band of TiO_2 are transported between TiO_2 nanoparticles with diffusion toward the back contact (TCO). And the electrons finally reach the counter electrode through the circuit.
- The oxidized photosensitizer (S^+) accepts electrons from the I^- ion redox mediator leading to regeneration of the ground state (S), and two I^--Ions are oxidized to elementary Iodine which reacts with I^- to the oxidized state, I_3^-.

  $$S^+ + e^- \rightarrow S$$

- The oxidized redox mediator, I_3^-, diffuses toward the counter electrode and then it is reduced to I^- ions.

  $$I_3^- + 2\,e^- \rightarrow 3\,I^-$$

The efficiency of a DSSC depends on four energy levels of the component: the excited state (approximately LUMO) and the ground state (HOMO) of the photosensitizer, the Fermi level of the TiO_2 electrode and the redox potential of the mediator (I^-/I_3^-) in the electrolyte.

Nanoplant-like Morphology

In DSSC, electrodes consisted of sintered semiconducting nanoparticles, mainly TiO_2 or ZnO. These nanoparticle DSSCs rely on trap-limited diffusion through the semiconductor nanoparticles for the electron transport. This limits the device efficiency since it is a slow transport mechanism. Recombination is more likely to occur at longer wavelengths of radiation. Moreover, sintering of nanoparticles requires a high temperature of about 450 °C, which restricts the fabrication of these cells to robust, rigid solid substrates. It has been proven that there is an increase in the efficiency of DSSC, if the sintered nanoparticle electrode is replaced by a specially designed electrode possessing an exotic 'nanoplant-like' morphology.

Operation

Sunlight enters the cell through the transparent SnO_2:F top contact, striking the dye on the surface of the TiO_2. Photons striking the dye with enough energy to be absorbed create an excited state of the dye, from which an electron can be "injected" directly into the conduction band of the TiO_2. From there it moves by diffusion (as a result of an electron concentration gradient) to the clear anode on top.

Meanwhile, the dye molecule has lost an electron and the molecule will decompose if another electron is not provided. The dye strips one from iodide in electrolyte below the TiO_2, oxidizing it into triiodide. This reaction occurs quite quickly compared to the time that it takes for the injected electron to recombine with the oxidized dye molecule, preventing this recombination reaction that would effectively short-circuit the solar cell.

The triiodide then recovers its missing electron by mechanically diffusing to the bottom of the cell, where the counter electrode re-introduces the electrons after flowing through the external circuit.

Efficiency

Several important measures are used to characterize solar cells. The most obvious is the total amount of electrical power produced for a given amount of solar power shining on the cell. Expressed as a percentage, this is known as the solar conversion efficiency. Electrical power is the product of current and voltage, so the maximum values for these measurements are important as well, J_{sc} and V_{oc} respectively. Finally, in order to understand the underlying physics, the "quantum efficiency" is used to compare the chance that one photon (of a particular energy) will create one electron.

In quantum efficiency terms, DSSCs are extremely efficient. Due to their "depth" in the nanostructure there is a very high chance that a photon will be absorbed, and the dyes are very effective at converting them to electrons. Most of the small losses that do exist in DSSC's are due to conduction losses in the TiO_2 and the clear electrode, or optical losses in the front electrode. The overall quantum efficiency for green light is about 90%, with the "lost" 10% being largely accounted for by the optical losses in the top electrode. The quantum efficiency of traditional designs vary, depending on their thickness, but are about the same as the DSSC.

In theory, the maximum voltage generated by such a cell is simply the difference between the (quasi-)Fermi level of the TiO_2 and the redox potential of the electrolyte, about 0.7 V under solar illumination conditions (V_{oc}). That is, if an illuminated DSSC is connected to a voltmeter in an "open circuit", it would read about 0.7 V. In terms of voltage, DSSCs offer slightly higher V_{oc} than silicon, about 0.7 V compared to 0.6 V. This is a fairly small difference, so real-world differences are dominated by current production, J_{sc}.

Although the dye is highly efficient at converting absorbed photons into free electrons in the TiO_2, only photons absorbed by the dye ultimately produce current. The rate of photon absorption depends upon the absorption spectrum of the sensitized TiO_2 layer and upon the solar flux spectrum. The overlap between these two spectra determines the maximum possible photocurrent. Typically used dye molecules generally have poorer absorption in the red part of the spectrum compared to silicon, which means that fewer of the photons in sunlight are usable for current generation. These factors limit the current generated by a DSSC, for comparison, a traditional silicon-based solar cell offers about 35 mA/cm^2, whereas current DSSCs offer about 20 mA/cm^2.

Overall peak power conversion efficiency for current DSSCs is about 11%. Current record for prototypes lies at 15%.

Degradation

DSSCs degrade when exposed to ultraviolet radiation. In 2014 air infiltration of the commonly-used amorphous Spiro-MeOTAD hole-transport layer was identified as the primary cause of the degradation, rather than oxidation. The damage could be avoided by the addition of an appropriate barrier.

The barrier layer may include UV stabilizers and/or UV absorbing luminescent chromophores (which emit at longer wavelengths which may be reabsorbed by the dye) and antioxidants to protect and improve the efficiency of the cell.

Advantages

DSSCs are currently the most efficient third-generation (2005 Basic Research Solar Energy

Utilization 16) solar technology available. Other thin-film technologies are typically between 5% and 13%, and traditional low-cost commercial silicon panels operate between 14% and 17%. This makes DSSCs attractive as a replacement for existing technologies in "low density" applications like rooftop solar collectors, where the mechanical robustness and light weight of the glass-less collector is a major advantage. They may not be as attractive for large-scale deployments where higher-cost higher-efficiency cells are more viable, but even small increases in the DSSC conversion efficiency might make them suitable for some of these roles as well.

There is another area where DSSCs are particularly attractive. The process of injecting an electron directly into the TiO_2 is qualitatively different from that occurring in a traditional cell, where the electron is "promoted" within the original crystal. In theory, given low rates of production, the high-energy electron in the silicon could re-combine with its own hole, giving off a photon (or other form of energy) which does not result in current being generated. Although this particular case may not be common, it is fairly easy for an electron generated by another atom to combine with a hole left behind in a previous photoexcitation.

In comparison, the injection process used in the DSSC does not introduce a hole in the TiO_2, only an extra electron. Although it is energetically possible for the electron to recombine back into the dye, the rate at which this occurs is quite slow compared to the rate that the dye regains an electron from the surrounding electrolyte. Recombination directly from the TiO_2 to species in the electrolyte is also possible although, again, for optimized devices this reaction is rather slow. On the contrary, electron transfer from the platinum coated electrode to species in the electrolyte is necessarily very fast.

As a result of these favorable "differential kinetics", DSSCs work even in low-light conditions. DSSCs are therefore able to work under cloudy skies and non-direct sunlight, whereas traditional designs would suffer a "cutout" at some lower limit of illumination, when charge carrier mobility is low and recombination becomes a major issue. The cutoff is so low they are even being proposed for indoor use, collecting energy for small devices from the lights in the house.

A practical advantage which DSSCs share with most thin-film technologies, is that the cell's mechanical robustness indirectly leads to higher efficiencies at higher temperatures. In any semiconductor, increasing temperature will promote some electrons into the conduction band "mechanically". The fragility of traditional silicon cells requires them to be protected from the elements, typically by encasing them in a glass box similar to a greenhouse, with a metal backing for strength. Such systems suffer noticeable decreases in efficiency as the cells heat up internally. DSSCs are normally built with only a thin layer of conductive plastic on the front layer, allowing them to radiate away heat much easier, and therefore operate at lower internal temperatures.

Disadvantages

The major disadvantage to the DSSC design is the use of the liquid electrolyte, which has temperature stability problems. At low temperatures the electrolyte can freeze, halting power production and potentially leading to physical damage. Higher temperatures cause the liquid to expand, making sealing the panels a serious problem. Another disadvantage is that costly ruthenium (dye), platinum (catalyst) and conducting glass or plastic (contact) are needed to produce a DSSC. A third major drawback is that the electrolyte solution contains volatile organic compounds (or VOC's), solvents which must be carefully sealed as they are hazardous to human health and the environment. This, along with the fact that the solvents permeate plastics, has precluded large-scale outdoor application and integration into flexible structure.

Replacing the liquid electrolyte with a solid has been a major ongoing field of research. Recent experiments using solidified melted salts have shown some promise, but currently suffer from higher degradation during continued operation, and are not flexible.

Photocathodes and Tandem Cells

Dye sensitised solar cells operate as a photoanode (n-DSC), where photocurrent result from electron injection by the sensitized dye. Photocathodes (p-DSCs) operate in an inverse mode compared to the conventional n-DSC, where dye-excitation is followed by rapid electron transfer from a p-type semiconductor to the dye (dye-sensitized hole injection, instead of electron injection). Such p-DSCs and n-DSCs can be combined to construct tandem solar cells (pn-DSCs) and the theoretical efficiency of tandem DSCs is well beyond that of single-junction DSCs.

A standard tandem cell consists of one n-DSC and one p-DSC in a simple sandwich configuration with an intermediate electrolyte layer. n-DSC and p-DSC are connected in series, which implies that the resulting photocurrent will be controlled by the weakest photoelectrode, whereas photovoltages are additive. Thus, photocurrent matching is very important for the construction of highly efficient tandem pn-DSCs. However, unlike n-DSCs, fast charge recombination following dye-sensitized hole injection usually resulted in low photocurrents in p-DSC and thus hampered the efficiency of the overall device.

Researchers have found that using dyes comprising a perylenemonoimide (PMI) as the acceptor and an oligothiophene coupled to triphenylamine as the donor greatly improve the performance of p-DSC by reducing charge recombination rate following dye-sensitized hole injection. The researchers constructed a tandem DSC device with NiO on the p-DSC side and TiO_2 on the n-DSC side. Photocurrent matching was achieved through adjustment of NiO and TiO_2 film thicknesses to control the optical absorptions and therefore match the photocurrents of both electrodes. The energy conversion efficiency of the device is 1.91%, which exceeds the efficiency of its individual components, but is still much lower than that of high performance n-DSC devices (6%–11%). The results are still promising since the tandem DSC was in itself rudimentary. The dramatic improvement in performance in p-DSC can eventually lead to tandem devices with much greater efficiency than lone n-DSCs.

Using electron transporting materials such as PCBM, TiO_2 and ZnO instead of the conventional liquid redox couple electrolyte, researchers have managed to fabricate the solid state p-DSC (p-ssDSCs), aiming for solid state tandem dye sensitized solar cells.

Development

The dyes used in early experimental cells were sensitive only in the high-frequen-cy end of the solar spectrum, in the UV and blue. Newer versions were quickly introduced that had much wider frequency response, notably the formula of triscarboxy-ruthenium terpyridine is

[Ru(4,4',4"-$(COOH)_3$-terpy)$(NCS)_3$], which is efficient right into the low-frequency range of red and IR light. The wide spectral response results in the dye having a deep brown-black color, and is referred to simply as "black dye". The dyes have an excellent chance of converting a photon into an electron, originally around 80% but improving to almost perfect conversion in more recent dyes, the over-all efficiency is about 90%, with the "lost" 10% being largely accounted for by the optical losses in top electrode.

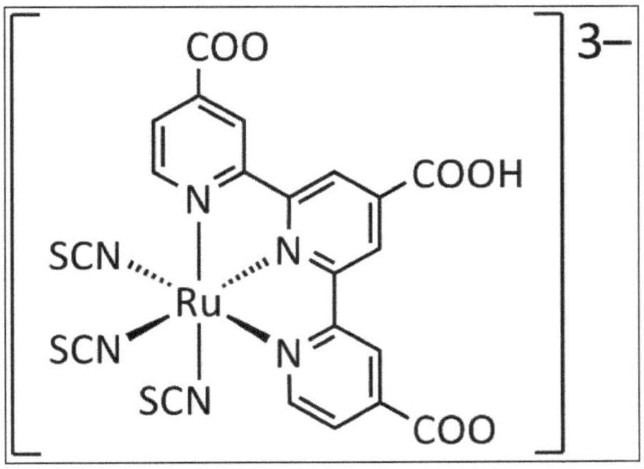

"Black Dye", an anionic Ru-terpyridine complex.

A solar cell must be capable of producing electricity for at least twenty years, without a significant decrease in efficiency (life span). The "black dye" system was subjected to 50 million cycles, the equivalent of ten years' exposure to the sun in Switzerland. No discernible performance decrease was observed. However the dye is subject to breakdown in high-light situations. Over the last decade an extensive research program has been carried out to address these concerns. The newer dyes included 1-ethyl-3 methylimidazolium tetrocyanoborate [EMIB$(CN)_4$] which is extremely light- and temperature-stable, copper-diselenium [Cu(In,GA)Se_2] which offers higher conversion efficiencies, and others with varying special-purpose properties.

DSSCs are still at the start of their development cycle. Efficiency gains are possible and have recently started more widespread study. These include the use of quantum dots for conversion of higher-energy (higher frequency) light into multiple electrons, using solid-state electrolytes for better temperature response, and changing the doping of the TiO_2 to better match it with the electrolyte being used.

Market Introduction

Several commercial providers are promising availability of DSCs in the near future:

- Dyesol officially opened its new manufacturing facilities in Queanbeyan Australia on 7 October 2008. It has subsequently announced partnerships with Tata Steel (TATA-Dyesol) and Pilkington Glass (Dyetec-Solar) for the development and large scale manufacture of DSC BIPV. Dyesol has also entered working relationships with Merck, Umicore, CSIRO, Japanese Ministry of Economy and Trade, Singapore Aerospace Manufacturing and a joint Venture with TIMO Korea (Dyesol-TIMO).
- Solaronix, a Swiss company specialized in the production of DSC materials since 1993, has extended their premises in 2010 to host a manufacturing pilot line of DSC modules.

- SolarPrint was founded in Ireland in 2008 by Dr. Mazhar Bari, Andre Fernon and Roy Horgan. SolarPrint was the first Ireland-based commercial entity involved in the manufacturing of PV technology. SolarPrint's innovation was the solution to the solvent-based electrolyte which to date has prohibited the mass commercialisation of DSSC. The company went into receivership in 2014 and was wound up.
- G24innovations founded in 2006, based in Cardiff, South Wales, UK. On 17 October 2007, claimed the production of the first commercial grade dye sensitised thin films.
- Sony Corporation has developed dye-sensitized solar cells with an energy conversion efficiency of 10%, a level seen as necessary for commercial use.
- Tasnee Enters Strategic Investment Agreement with Dyesol.
- H.Glass was founded 2011 in Switzerland. H.Glass has put enormous efforts to create industrial process for the DSSC technologie - the first results where shown at the EXPO 2015 in Milano at the Austrian Pavilion. The milestone for DSSC is the Science Tower in Austria - it is the largest installation of DSSC in the world - carried out by SFL technologies.

ORGANIC SOLAR CELL

Organic photovoltaic (OPV) solar cells aim to provide an Earth-abundant and low-energy-production photovoltaic (PV) solution. This technology also has the theoretical potential to provide electricity at a lower cost than first- and second-generation solar technologies. Because various absorbers can be used to create colored or transparent OPV devices, this technology is particularly appealing to the building-integrated PV market. Organic photovoltaics have achieved efficiencies near 11%, but efficiency limitations as well as long-term reliability remain significant barriers.

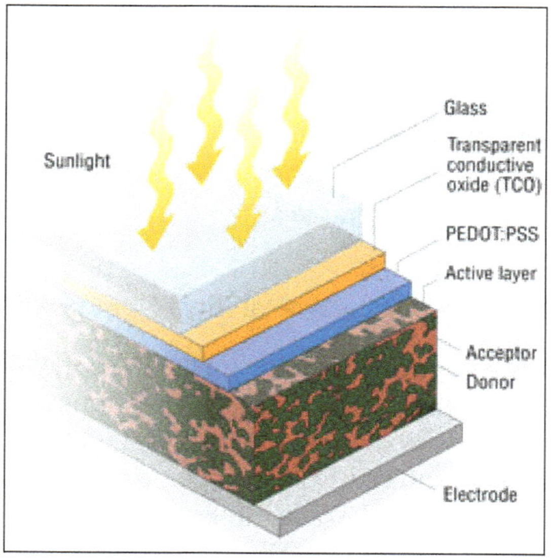

Organic photovoltaic cell structure.

Unlike most inorganic solar cells, OPV cells use molecular or polymeric absorbers, which results in a localized exciton. The absorber is used in conjunction with an electron acceptor, such as a fullerene, which has molecular orbital energy states that facilitate electron transfer. Upon absorbing a photon, the resulting exciton migrates to the interface between the absorber material and the electron acceptor material. At the interface, the energetic mismatch of the molecular orbitals provides sufficient driving force to split the exciton and create free charge carriers (an electron and a hole).

Research Directions

The low efficiencies of OPV cells are related to their small exciton diffusion lengths and low carrier mobilities. These two characteristics ultimately result in the use of thin active layers that affect overall device performance. Furthermore, the operational lifetime of OPV modules remains significantly lower than for inorganic devices.

Current research focuses on increasing device efficiency and lifetime. Substantial efficiency gains have been achieved already by improving the absorber material, and research is being done to further optimize the absorbers and develop an organic multijunction architectures. Improved encapsulation and alternative contact materials are being investigated to reduce cell degradation and push cell lifetimes to industry-relevant values.

Benefits

The benefits promised by OPV solar cells include:

- Low-cost manufacturing: Soluble organic molecules enable roll-to-roll processing techniques and allow for low-cost manufacturing.
- Abundant materials: The wide abundance of building-block materials may reduce supply and price constraints.
- Flexible substrates: The ability to be applied to flexible substrates permits a wide variety of uses.

Production

OPV cells are categorized into two classes:

- Small-molecule OPV cells
- Polymer-based OPV cells

Small-molecule OPV cells use molecules with broad absorption in the visible and near-infrared portion of the electromagnetic spectrum. Highly conjugated systems are typically used for the electron-donating system such as phthalocyanines, polyacenes, and squarenes. Perylene dyes and fullerenes are often used as the electron-accepting systems. These devices are most commonly generated via vacuum deposition to create bilayer and tandem architectures. Recently, solution-processed small-molecule systems have been developed.

Polymer-based OPV cells use long-chained molecular systems for the electron-donating material (e.g., P3HT, MDMO-PPV), along with derivatized fullerenes as the electron-accepting system (e.g., $PC_{60}BM$, $PC_{70}BM$). Like small-molecule OPV cells, these systems have small exciton diffusion lengths. However, this limitation is circumvented by a high interface surface area within the active device.

Dye-sensitized solar cells are a hybrid organic-inorganic technology that uses small-molecule absorber dyes. These dyes adsorb onto a suitable electron-accepting material, such as titanium dioxide or zinc oxide, along with an electrolyte to regenerate the dye.

HYBRID SOLAR CELL

Hybrid solar cells combine advantages of both organic and inorganic semiconductors. Hybrid photovoltaics have organic materials that consist of conjugated polymers that absorb light as the donor and transport holes. Inorganic materials in hybrid cells are used as the acceptor and electron transporter in the structure. The hybrid photovoltaic devices have a potential for not only low-cost by roll-to-roll processing but also for scalable solar power conversion.

Theory

Solar cells are devices that convert sunlight into electricity by the photovoltaic effect. Electrons in a solar cell absorb photon energy in sunlight which excites them to the conduction band from the valence band. This generates a hole-electron pair, which is separated by a potential barrier (such as a p-n junction), and induces a current. Organic solar cells use organic materials in their active layers. Molecular, polymer, and hybrid organic photovoltaics are the main kinds of organic photovoltaic devices currently studied.

Hybrid Solar Cell

In hybrid solar cells, an organic material is mixed with a high electron transport material to form the photoactive layer. The two materials are assembled together in a heterojunction-type photoactive layer, which can have a greater power conversion efficiency than a single material. One of the materials acts as the photon absorber and exciton donor. The other material facilitates exciton dissociation at the junction. Charge is transferred and then separated after an exciton created in the donor is delocalized on a donor-acceptor complex.

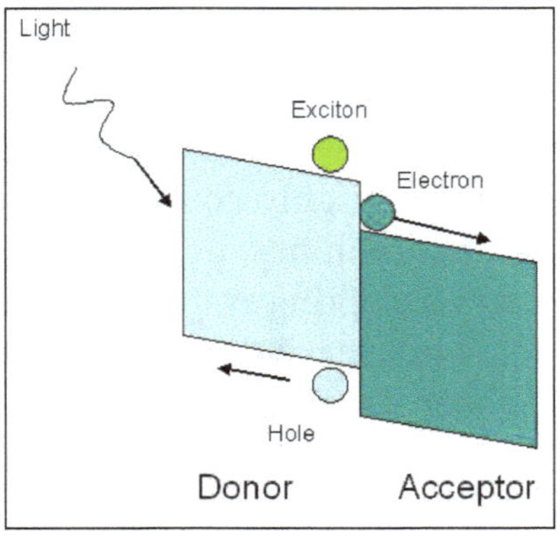

Energy diagram of the donor and acceptor. The conduction band of the acceptor is lower than the LUMO of the polymer, allowing for transfer of the electron.

The acceptor material needs a suitable energy offset to the binding energy of the exciton to the absorber. Charge transfer is favorable if the following condition is satisfied:

$$E_A^A - E_A^D > U_D$$

where superscripts A and D refer to the acceptor and donor respectively, E_A is the electron affinity, and U the coulombic binding energy of the exciton on the donor. An energy diagram of the interface is shown in figure. In commonly used photovoltaic polymers such as MEH-PPV, the exciton binding energy ranges from 0.3 eV to 1.4 eV.

The energy required to separate the exciton is provided by the energy offset between the LUMOs or conduction bands of the donor and acceptor. After dissociation, the carriers are transported to the respective electrodes through a percolation network.

The average distance an exciton can diffuse through a material before annihilation by recombination is the exciton diffusion length. This is short in polymers, on the order of 5–10 nanometers. The time scale for radiative and non-radiative decay is from 1 picosecond to 1 nanosecond. Excitons generated within this length close to an acceptor would contribute to the photocurrent.

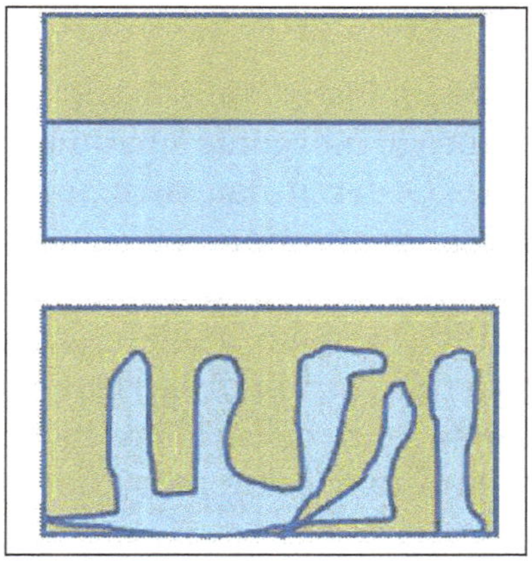

Two different structures of heterojunctions, a) phase separated bi-layer and b) bulk heterojunction. The bulk heterojunction allows for more interfacial contact between the two phases, which is beneficial for the nanoparticle-polymer compound as it provides more surface area for charge transfer.

To deal with the problem of the short exciton diffusion length, a bulk heterojunction structure is used rather than a phase-separated bilayer. Dispersing the particles throughout the polymer matrix creates a larger interfacial area for charge transfer to occur. Figure displays the difference between a bilayer and a bulk heterojunction.

Types of Interfaces and Structures

Controlling the interface of inorganic-organic hybrid solar cells can increase the efficiency of the cells. This increased efficiency can be achieved by increasing the interfacial surface area between the organic and the inorganic to facilitate charge separation and by controlling the nanoscale lengths and periodicity of each structure so that charges are allowed to separate and move toward the appropriate electrode without recombining. The three main nanoscale structures used are mesoporous inorganic films infused with electron-donating organic, alternatining inorganic-organic lamellar structures, and nanowire structures.

Mesoporous Films

Mesoporous films have been used for a relatively high-efficiency hybrid solar cell. The structure of mesoporous thin film solar cells usually includes a porous inorganic that is saturated with organic surfactant. The organic absorbs light, and transfers electrons to the inorganic semiconductor (usually a transparent conducting oxide), which then transfers the electron to the electrode. Problems with these cells include their random ordering and the difficulty of controlling their nanoscale structure to promote charge conduction.

Ordered Lamellar Films

Recently, the use of alternating layers of organic and inorganic compounds has been controlled through electrodeposition-based self-assembly. This is of particular interest because it has been shown that the lamellar structure and periodicity of the alternating organic-inorganic layers can be controlled through solution chemistry. To produce this type of cell with practical efficiencies, larger organic surfactants that absorb more of the visible spectrum must be deposited between the layers of electron-accepting inorganic.

Films of Ordered Nanostructures

Researchers have been able to grow nanostructure-based solar cells that use ordered nanostructures like nanowires or nanotubes of inorganic surrounding by electron-donating organics utilizing self-organization processes. Ordered nanostructures offer the advantage of directed charge transport and controlled phase separation between donor and acceptor materials. The nanowire-based morphology offers reduced internal reflection, facile strain relaxation and increased defect tolerance. The ability to make single-crystalline nanowires on low-cost substrates such as aluminum foil and to relax strain in subsequent layers removes two more major cost hurdles associated with high-efficiency cells. There have been rapid increases in efficiencies of nanowire-based solar cells and they seem to be one of the most promising nanoscale solar hybrid technologies.

Fundamental Challenge Factors

Hybrid cell efficiency must be increased to start large-scale manufacturing. Three factors affect efficiency. First, the bandgap should be reduced to absorb red photons, which contain a significant fraction of the energy in the solar spectrum. Current organic photovoltaics have shown 70% of quantum efficiency for blue photons. Second, contact resistance between each layer in the device should be minimized to offer higher fill factor and power conversion efficiency. Third, charge-carrier mobility should be increased to allow the photovoltaics to have thicker active layers while minimizing carrier recombination and keeping the series resistance of the device low.

Types of Hybrid Solar Cells

Polymer: Nanoparticle Composite

Nanoparticles are a class of semiconductor materials whose size in at least one dimension ranges from 1 to 100 nanometers, on the order of exciton wavelengths. This size control creates quantum confinement and allows for the tuning of optoelectronic properties, such as band gap and electron affinity. Nanoparticles also have a large surface area to volume ratio, which presents more area for charge transfer to occur.

The photoactive layer can be created by mixing nanoparticles into a polymer matrix. Solar devices based on polymer-nanoparticle composites most resemble polymer solar cells. In this case, the nanoparticles take the place of the fullerene based acceptors used in fully organic polymer solar cells. Hybrid solar cells based upon nanoparticles are an area of research interest because nanoparticles have several properties that could make them preferable to fullerenes, such as:

- Fullerenes are synthesized by a combination of a high temperature arc method and continuous gas-phase synthesis, which makes their production difficult and energy intensive. The colloidal synthesis of nanoparticles by contrast is a low temperature process.
- PCBM (a common fullerene acceptor) diffuses during long timespans or when exposed to heat, which can alter the morphology and lower the efficiency of a polymer solar cell. Limited testing of nanoparticle solar cells indicates they may be more stable over time.
- Nanoparticles are more absorbent than fullerenes, meaning more light can be theoretically absorbed in a thinner device.
- Nanoparticle size can affect absorption. This combined with the fact that there are many possible semiconducting nanoparticles allows for highly customizable bandgaps that can be easily tuned to certain frequencies, which would be advantageous in tandem solar cells.
- Nanoparticles with size near their Bohr radius can generate two excitons when struck by a sufficiently energetic photon.

Structure and Processing

For polymers used in this device, hole mobilities are greater than electron mobilities, so the polymer phase is used to transport holes. The nanoparticles transport electrons to the electrode.

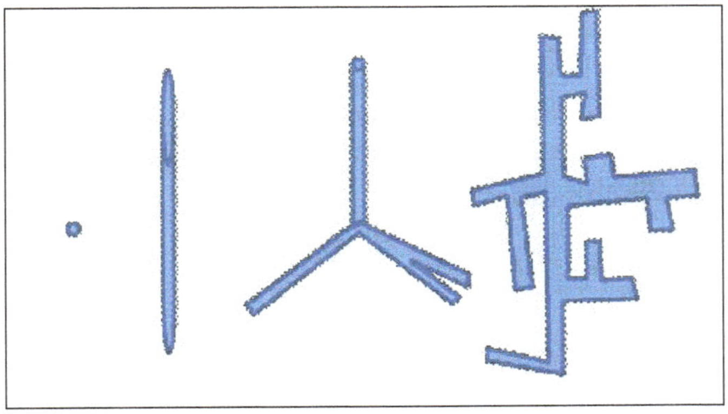

Four different structures of nanoparticles, which have at least 1 dimension in the 1–100 nm range, retaining quantum confinement. Left is a nanocrystal, next to it is nanorod, third is tetrapod, and right is hyperbranched.

The interfacial area between the polymer phase and the nanoparticles needs to be large. This is achieved by dispersing the particles throughout the polymer matrix. However, the nanoparticles need to be interconnected to form percolation networks for electron transport, which occurs by hopping events.

Efficiency is affected by aspect ratio, geometry, and volume fraction of the nanoparticles. Nanoparticle structures include nanocrystals, nanorods, and hyperbranched structures. Figure contains a picture of each structure. Different structures change the conversion efficiency by effecting nanoparticle dispersion in the polymer and providing pathways for electron transport.

The nanoparticle phase is required to provide a pathway for the electrons to reach the electrode. By using nanorods instead of nanocrystals, the hopping event from one crystal to another can be avoided.

Fabrication methods include mixing the two materials in a solution and spin-coating it onto a substrate, and solvent evaporation (sol-gel). Most of these methods do not involve high-temperature processing. Annealing increases order in the polymer phase, increasing conductivity. However, annealing for too long causes the polymer domain size to increase, eventually making it larger than the exciton diffusion length, and possibly allowing some of the metal from the contact to diffuse into the photoactive layer, reducing the efficiency of the device.

Materials

Inorganic semiconductor nanoparticles used in hybrid cells include CdSe (size ranges from 6–20 nm), ZnO, TiO, and PbS. Common polymers used as photo materials have extensive conjugation and are also hydrophobic. Their efficiency as a photo-material is affected by the HOMO level position and the ionization potential, which directly affects the open circuit voltage and the stability in air. The most common polymers used are P3HT (poly (3-hexylthiophene)), and M3H-PPV (poly[2-methoxy, 5-(2′-ethyl-hexyloxy)-p-phenylenevinylene)]). P3HT has a bandgap of 2.1 eV and M3H-PPV has a bandgap of ~2.4 eV. These values correspond with the bandgap of CdSe, 2.10 eV. The electron affinity of CdSe ranges from 4.4 to 4.7 eV. When the polymer used is MEH-PPV, which has an electron affinity of 3.0 eV, the difference between the electron affinities is large enough to drive electron transfer from the CdSe to the polymer. CdSe also has a high electron mobility (600 $cm^2 \cdot V^{-1} \cdot s^{-1}$).

Performance Values

The highest demonstrated efficiency is 3.2%, based upon a PCPDTBT polymer donor and CdSe nanoparticle acceptor. The device exhibited a short circuit current of 10.1 $mA \cdot cm^{-2}$, an open circuit voltage of .68 V, and a fill factor of .51.

Challenges

Hybrid solar cells need increased efficiencies and stability over time before commercialization is feasible. In comparison to the 2.4% of the CdSe-PPV system, silicon photodevices have power conversion efficiencies greater than 20%.

Problems include controlling the amount of nanoparticle aggregation as the photolayer forms. The particles need to be dispersed in order to maximize interface area, but need to aggregate to form networks for electron transport. The network formation is sensitive to the fabrication conditions.

Dead end pathways can impede flow. A possible solution is implementing ordered heterojunctions, where the structure is well controlled.

The structures can undergo morphological changes over time, namely phase separation. Eventually, the polymer domain size will be greater than the carrier diffusion length, which lowers performance.

Even though the nanoparticle bandgap can be tuned, it needs to be matched with the corresponding polymer. The 2.0 eV bandgap of CdSe is larger than an ideal bandgap of 1.4 for absorbance of light.

The nanoparticles involved are typically colloids, which are stabilized in solution by ligands. The ligands decrease device efficiency because they serve as insulators which impede interaction between the donor and nanoparticle acceptor as well as decreasing the electron mobility. Some, but not complete success has been had by exchanging the initial ligands for pyridine or another short chain ligand.

Hybrid solar cells exhibit material properties inferior to those of bulk silicon semiconductors. The carrier mobilities are much smaller than that of silicon. Electron mobility in silicon is 1000 $cm^2 \cdot V^{-1} \cdot s^{-1}$, compared to 600 $cm^2 \cdot V^{-1} \cdot s^{-1}$ in CdSe, and less than 10 $cm^2 \cdot V^{-1} \cdot s^{-1}$ in other quantum dot materials. Hole mobility in MEH-PPV is 0.1 $cm^2 \cdot V^{-1} \cdot s^{-1}$, while in silicon it is 450 $cm^2 \cdot V^{-1} \cdot s^{-1}$.

Carbon Nanotubes

Carbon nanotubes (CNTs) have high electron conductivity, high thermal conductivity, robustness, and flexibility. Field emission displays (FED), strain sensors, and field effect transistors (FET) using CNTs have been demonstrated. Each application shows the potential of CNTs for nanoscale devices and for flexible electronics applications. Photovoltaic applications have also been explored for this material.

Mainly, CNTs have been used as either the photo-induced exciton carrier transport medium impurity within a polymer-based photovoltaic layer or as the photoactive (photon-electron conversion) layer. Metallic CNT is preferred for the former application, while semiconducting CNT is preferred for the later.

Efficient Carrier Transport Medium

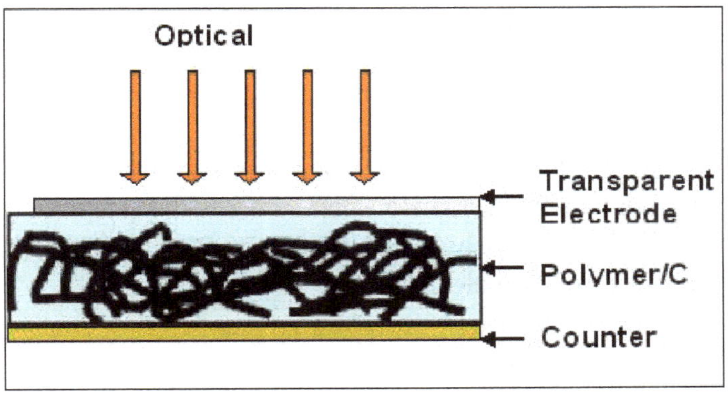

Device diagram for CNT as efficient carrier transport medium.

To increase the photovoltaic efficiency, electron-accepting impurities must be added to the photoactive region. By incorporating CNTs into the polymer, dissociation of the exciton pair can be accomplished by the CNT matrix. The high surface area (~1600 m^2/g) of CNTs offers a good opportunity for exciton dissociation. The separated carriers within the polymer-CNT matrix are transported by the percolation pathways of adjacent CNTs, providing the means for high carrier mobility and efficient charge transfer. The factors of performance of CNT-polymer hybrid photovoltaics are low compared to those of inorganic photovoltaics. SWNT in P3OT semiconductor polymer demonstrated open circuit voltage (V_{oc}) of below 0.94 V, with short circuit current (I_{sc}) of 0.12 mA/cm^2.

Metal nanoparticles may be applied to the exterior of CNTs to increase the exciton separation efficiency. The metal provides a higher electric field at the CNT-polymer interface, accelerating the exciton carriers to transfer them more effectively to the CNT matrix. In this case, V_{oc} = 0.3396 V and I_{sc} = 5.88 mA/cm^2. The fill factor is 0.3876%, and the white light conversion factor 0.775%.

Photoactive Matrix Layer

CNT may be used as a photovoltaic device not only as an add-in material to increase carrier transport, but also as the photoactive layer itself. The semiconducting single walled CNT (SWCNT) is a potentially attractive material for photovoltaic applications for the unique structural and electrical properties. SWCNT has high electric conductivity (100 times that of copper) and shows ballistic carrier transport, greatly decreasing carrier recombination. The bandgap of the SWCNT is inversely proportional to the tube diameter, which means that SWCNT may show multiple direct bandgaps matching the solar spectrum.

A strong built-in electric field in SWCNT for efficient photogenerated electron-hole pair separation has been demonstrated by using two asymmetrical metal electrodes with high and low work functions. The open circuit voltage (V_{oc}) is 0.28 V, with short circuit current (I_{sc}) 1.12 $nA \cdot cm^{-2}$ with an incident light source of 8.8 $W \cdot cm^{-2}$. The resulting white light conversion factor is 0.8%.

Challenges

Several challenges must be addressed for CNT to be used in photovoltaic applications. CNT degrades over time in an oxygen-rich environment. The passivation layer required to prevent CNT oxidation may reduce the optical transparency of the electrode region and lower the photovoltaic efficiency.

Challenges as Efficient Carrier Transport Medium

Additional challenges involve the dispersion of CNT within the polymer photoactive layer. The CNT is required to be well dispersed within the polymer matrix to form charge-transfer-efficient pathways between the excitons and the electrode.

Challenges as Photoactive Matrix Layer

Challenges of CNT for the photoactive layer include its lack of capability to form a p-n junction, due to the difficulty of doping certain segments of a CNT. (A p-n junction creates an internal built-in potential, providing a pathway for efficient carrier separation within the photovoltaic.) To overcome this difficulty, energy band bending has been done by the use of two electrodes of different work functions. A strong built-in electric field covering the whole SWCNT channel is formed for high-efficiency carrier separation. The oxidation issue with CNT is more critical for this application. Oxidized CNTs have a tendency to become more metallic, and so less useful as a photovoltaic material.

Dye-sensitized

Dye-sensitized solar cells consists of a photo-sensitized anode, an electrolyte, and a photo-electrochemical system. Hybrid solar cells based on dye-sensitized solar cells are formed with inorganic materials (TiO_2) and organic materials.

Materials

Hybrid solar cells based on dye-sensitized solar cells are fabricated by dye-absorbed inorganic materials and organic materials. TiO_2 is the preferred inorganic material since this material is easy to synthesize and acts as a n-type semiconductor due to the donor-like oxygen vacancies. However, titania only absorbs a small fraction of the UV spectrum. Molecular sensitizers (dye molecules) attached to the semiconductor surface are used to collect a greater portion of the spectrum. In the case of titania dye-sensitized solar cells, a photon absorbed by a dye-sensitizer molecule layer induces electron injection into the conduction band of titania, resulting in current flow. However, short diffusion length (diffusivity, $D_n \leq 10^{-4} cm^2/s$) in titania dye-sensitized solar cells decrease the solar-to-energy conversion efficiency. To enhance diffusion length (or carrier lifetime), a variety of organic materials are attached to the titania.

Fabrication Scheme

Dye-sensitized Photoelectrochemical Cell (Grätzel Cell)

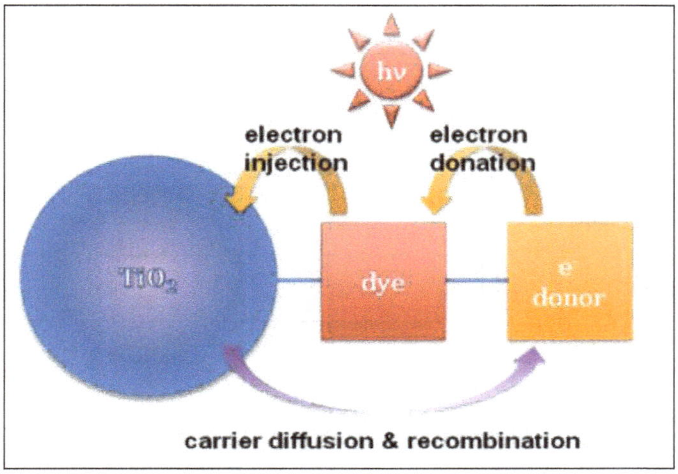

Schematic representation of electron-hole generation and recombination.

TiO_2 nanoparticles are synthesized in several tens of nanometer scales (~100 nm). In order to make a photovoltaic cell, molecular sensitizers (dye molecules) are attached to the titania surface. The dye-absorbed titania is finally enclosed by a liquid electrolyte. This type of dye-sensitized solar cell is also known as a Grätzel cell. Dye-sensitized solar cell has a disadvantage of a short diffusion length. Recently, supermolecular or multifunctional sensitizers have been investigated so as to enhance carrier diffusion length. For example, a dye chromophore has been modified by the addition of secondary electron donors. Minority carriers (holes in this case) diffuse to the attached electron donors to recombine. Therefore, electron-hole recombination is retarded by the physical separation between the dye–cation moiety and the TiO_2 surface, as shown in figure. Finally, this process raises the carrier diffusion length, resulting in the increase of carrier lifetime.

Solid-state Dye Sensitized Solar Cell

Mesoporous materials contain pores with diameters between 2 and 50 nm. A dye-sensitized mesoporous film of TiO_2 can be used for making photovoltaic cells and this solar cell is called a 'solid-state dye sensitized solar cell'. The pores in mesoporous TiO_2 thin film are filled with a solid hole-conducting material such as p-type semiconductors or organic hole conducting material. Replacing the liquid electrolyte in Grätzel's cells with a solid charge-transport material can be beneficial. The process of electron-hole generation and recombination is the same as Grätzel cells. Electrons are injected from photoexcited dye into the conduction band of titania and holes are transported by a solid charge transport electrolyte to an electrode. Many organic materials have been tested to obtain a high solar-to-energy conversion efficiency in dye synthesized solar cells based on mesoporous titania thin film.

Efficiency Factors

Efficiency factors demonstrated for dye-sensitized solar cells are:

Parameters	Types of dye sensitized solar cells	
	Grätzel cell	Solid-state
Efficiency (%)	~ 10–11	~ 4
V_{oc} (V)	~ 0.7	~ 0.40
J_{sc} (mA/cm²)	~ 20	~ 9.10
Fill factor	~ 0.67	~ 0.6

Challenges

Liquid organic electrolytes contain highly corrosive iodine, leading to problems of leakage, sealing, handling, dye desorption, and maintenance. Much attention is now focused on the electrolyte to address these problems.

For solid-state dye sensitized solar cells, the first challenge originates from disordered titania mesoporous structures. Mesoporous titania structures should be fabricated with well-ordered titania structures of uniform size (~ 10 nm). The second challenge comes from developing the solid electrolyte, which is required to have these properties:

- The electrolyte should be transparent to the visible spectrum (wide band gap).
- Fabrication should be possible for depositing the solid electrolyte without degrading the dye molecule layer on titania.
- The LUMO of the dye molecule should be higher than the conduction band of titania.
- Several p-type semiconductors tend to crystallize inside the mesoporous titania films, destroying the dye molecule-titania contact. Therefore, the solid electrolyte needs to be stable during operation.

Nanostructured Inorganic — Small Molecules

In 2008, scientists were able to create a nanostructured lamellar structure that provides an ideal design for bulk heterojunction solar cells. The observed structure is composed of ZnO and small, conducting organic molecules, which co-assemble into alternating layers of organic and inorganic components. This highly organized structure, which is stabilized by π-π stacking between the organic molecules, allows for conducting pathways in both the organic and inorganic layers. The thicknesses of the layers (about 1–3 nm) are well within the exciton diffusion length, which ideally minimizes recombination among charge carriers. This structure also maximizes the interface between the inorganic ZnO and the organic molecules, which enables a high chromophore loading density within the structure. Due to the choice of materials, this system is non-toxic and environmentally friendly, unlike many other systems which use lead or cadmium.

Although this system has not yet been incorporated into a photovoltaic device, preliminary photoconductivity measurements have shown that this system exhibits among the highest values measured for organic, hybrid, and amorphous silicon photoconductors, and so, offers promise in creating efficient hybrid photovoltaic devices.

QUANTUM DOT SOLAR CELL

A quantum dot solar cell (QDSC) is a solar cell design that uses quantum dots as the absorbing photovoltaic material. It attempts to replace bulk materials such as silicon, copper indium gallium selenide (CIGS) or cadmium telluride (CdTe). Quantum dots have bandgaps that are tunable across a wide range of energy levels by changing their size. In bulk materials, the bandgap is fixed by the choice of material(s). This property makes quantum dots attractive for multi-junction solar cells, where a variety of materials are used to improve efficiency by harvesting multiple portions of the solar spectrum.

As of 2016, efficiency exceeds 10%.

The idea of using quantum dots as a path to high efficiency was first noted by Burnham and Duggan in 1990. At the time, the science of quantum dots, or "wells" as they were known, was in its infancy and early examples were just becoming available.

Spin-cast quantum dot solar cell built by the Sargent Group at the University of Toronto. The metal disks on the front surface are the electrical connections to the layers below.

DSSC Efforts

Another modern cell design is the dye-sensitized solar cell, or DSSC. DSSCs use a sponge-like layer of TiO_2 as the semiconductor valve as well as a mechanical support structure. During construction, the sponge is filled with an organic dye, typically ruthenium-polypyridine, which injects electrons into the titanium dioxide upon photoexcitation. This dye is relatively expensive, and ruthenium is a rare metal.

Using quantum dots as an alternative to molecular dyes was considered from the earliest days of DSSC research. The ability to tune the bandgap allowed the designer to select a wider variety of materials for other portions of the cell. Collaborating groups from the University of Toronto and École Polytechnique Fédérale de Lausanne developed a design based on a rear electrode directly in contact with a film of quantum dots, eliminating the electrolyte and forming a depleted heterojunction. These cells reached 7.0% efficiency, better than the best solid-state DSSC devices, but below those based on liquid electrolytes.

Multi-junction

Traditionally, multi-junction solar cells are made with a collection of multiple semiconductor materials. Because each material has a different band gap, each material's p-n junction will be optimized for a different incoming wavelength of light. Using multiple materials enables the absorbance of a broader range of wavelengths, which increases the cell's electrical conversion efficiency.

However, the use of multiple materials makes multi-junction solar cells too expensive for many commercial uses. Because the band gap of quantum dots can be tuned by adjusting the particle radius, multi-junction cells can be manufactured by incorporating quantum dot semiconductors of different sizes (and therefore different band gaps). Using the same material lowers manufacturing costs, and the enhanced absorption spectrum of quantum dots can be used to increase the short-circuit current and overall cell efficiency.

Cadmium telluride (CdTe) is used for cells that absorb multiple frequencies. A colloidal suspension of these crystals is spin-cast onto a substrate such as a thin glass slide, potted in a conductive polymer. These cells did not use quantum dots, but shared features with them, such as spin-casting and the use of a thin film conductor. At low production scales quantum dots are more expensive than mass-produced nanocrystals, but cadmium and telluride are rare and highly toxic metals subject to price swings.

The Sargent Group used lead sulfide as an infrared-sensitive electron donor to produce then record-efficiency IR solar cells. Spin-casting may allow the construction of "tandem" cells at greatly reduced cost. The original cells used a gold substrate as an electrode, although nickel works just as well.

Hot-carrier Capture

Another way to improve efficiency is to capture the extra energy in the electron when emitted from a single-bandgap material. In traditional materials like silicon, the distance from the emission site to the electrode where they are harvested is too far to allow this to occur; the electron will undergo many interactions with the crystal materials and lattice, giving up this extra energy as heat. Amorphous thin-film silicon was tried as an alternative, but the defects inherent to these materials overwhelmed their potential advantage. Modern thin-film cells remain generally less efficient than traditional silicon.

Nanostructured donors can be cast as uniform films that avoid the problems with defects. These would be subject to other issues inherent to quantum dots, notably resistivity issues and heat retention.

Multiple Excitons

The Shockley-Queisser limit, which sets the maximum efficiency of a single-layer photovoltaic cell to be 33.7%, assumes that only one electron-hole pair (exciton) can be generated per incoming photon. Multiple exciton generation (MEG) is an exciton relaxation pathway which allows two or more excitons to be generated per incoming high energy photon. In traditional photovoltaics, this excess energy is lost to the bulk material as lattice vibrations (electron-phonon coupling). MEG occurs when this excess energy is transferred to excite additional electrons across the band gap, where they can contribute to the short-circuit current density.

Within quantum dots, quantum confinement increases coulombic interactions which drives the MEG process. This phenomenon also decreases the rate of electron-phonon coupling, which is the dominant method of exciton relaxation in bulk semiconductors. The phonon bottleneck slows the rate of hot carrier cooling, which allows excitons to pursue other pathways of relaxation; this allows MEG to dominate in quantum dot solar cells. The rate of MEG can be optimized by tailoring quantum dot ligand chemistry, as well as by changing the quantum dot material and geometry.

In 2004, Los Alamos National Laboratory reported spectroscopic evidence that several excitons could be efficiently generated upon absorption of a single, energetic photon in a quantum dot. Capturing them would catch more of the energy in sunlight. In this approach, known as "carrier multiplication" (CM) or "multiple exciton generation" (MEG), the quantum dot is tuned to release multiple electron-hole pairs at a lower energy instead of one pair at high energy. This increases efficiency through increased photocurrent. LANL's dots were made from lead selenide.

In 2010, the University of Wyoming demonstrated similar performance using DCCS cells. Lead-sulfur (PbS) dots demonstrated two-electron ejection when the incoming photons had about three times the bandgap energy.

In 2005, NREL demonstrated MEG in quantum dots, producing three electrons per photon and a theoretical efficiency of 65%. In 2007, they achieved a similar result in silicon.

Non-oxidizing

In 2014 a University of Toronto group manufactured and demonstrated a type of CQD n-type cell using PbS with special treatment so that it doesn't bind with oxygen. The cell achieved 8% efficiency, just shy of the current QD efficiency record. Such cells create the possibility of uncoated "spray-on" cells. However, these air-stable n-type CQD were actually fabricated in an oxygen-free environment.

Also in 2014, another research group at MIT demonstrated air-stable ZnO/PbS solar cells that were fabricated in air and achieved a certified 8.55% record efficiency (9.2% in lab) because they absorbed light well, while also transporting charge to collectors at the cell's edge. These cells show unprecedented air-stability for quantum dot solar cells that the performance remained unchanged for more than 150 days of storage in air.

Solar Cell Concepts

In a conventional solar cell, light is absorbed by a semiconductor, producing an electron-hole (e-h) pair; the pair may be bound and is referred to as an exciton. This pair is separated by an internal electrochemical potential (present in p-n junctions or Schottky diodes) and the resulting flow of electrons and holes creates electric current. The internal electrochemical potential is created by doping one part of semiconductor interface with atoms that act as electron donors (n-type doping) and another with electron acceptors (p-type doping) that results in a p-n junction. Generation of an e-h pair requires that the photons have energy exceeding the bandgap of the material. Effectively, photons with energies lower than the bandgap do not get absorbed, while those that are higher can quickly (within about 10^{-13} s) thermalize to the band edges, reducing output. The former limitation reduces current, while the thermalization reduces the voltage. As a result, semiconductor cells suffer a trade-off between voltage and current (which can be in part alleviated by using multiple junction implementations). The detailed balance calculation shows that this efficiency can not exceed 33% if one uses a single material with an ideal bandgap of 1.34 eV for a solar cell.

The band gap (1.34 eV) of an ideal single-junction cell is close to that of silicon (1.1 eV), one of the many reasons that silicon dominates the market. However, silicon's efficiency is limited to about 30% (Shockley–Queisser limit). It is possible to improve on a single-junction cell by vertically stacking cells with different bandgaps – termed a "tandem" or "multi-junction" approach. The same analysis shows that a two layer cell should have one layer tuned to 1.64 eV and the other to 0.94 eV, providing a theoretical performance of 44%. A three-layer cell should be tuned to 1.83, 1.16 and 0.71 eV, with an efficiency of 48%. An "infinity-layer" cell would have a theoretical efficiency of 86%, with other thermodynamic loss mechanisms accounting for the rest.

Traditional (crystalline) silicon preparation methods do not lend themselves to this approach due to lack of bandgap tunability. Thin-films of amorphous silicon, which due to a relaxed requirement in crystal momentum preservation can achieve direct bandgaps and intermixing of carbon, can tune the bandgap, but other issues have prevented these from matching the performance of traditional cells. Most tandem-cell structures are based on higher performance semiconductors, notably indium gallium arsenide (InGaAs). Three-layer InGaAs/GaAs/InGaP cells (bandgaps 0.94/1.42/1.89 eV) hold the efficiency record of 42.3% for experimental examples.

However, the QDSCs suffer from weak absorption and the contribution of the light absorption at room temperature is marginal. This can be addressed by utilizing multibranched Au nanostars.

Quantum Dots

Quantum dots are semiconducting particles that have been reduced below the size of the Exciton Bohr radius and due to quantum mechanics considerations, the electron energies that can exist within them become finite, much alike energies in an atom. Quantum dots have been referred to as "artificial atoms". These energy levels are tuneable by changing their size, which in turn defines the bandgap. The dots can be grown over a range of sizes, allowing them to express a variety of bandgaps without changing the underlying material or construction techniques. In typical wet chemistry preparations, the tuning is accomplished by varying the synthesis duration or temperature.

The ability to tune the bandgap makes quantum dots desirable for solar cells. For the sun's photon distribution spectrum, the Shockley-Queisser limit indicates that the maximum solar conversion efficiency occurs in a material with a band gap of 1.34 eV. However, materials with lower band gaps will be better suited to generate electricity from lower-energy photons (and vice versa). Single junction implementations using lead sulfide (PbS) colloidal quantum dots (CQD) have bandgaps that can be tuned into the far infrared, frequencies that are typically difficult to achieve with traditional solar cells. Half of the solar energy reaching the Earth is in the infrared, most in the near infrared region. A quantum dot solar cell makes infrared energy as accessible as any other.

Moreover, CQD offer easy synthesis and preparation. While suspended in a colloidal liquid form they can be easily handled throughout production, with a fumehood as the most complex equipment needed. CQD are typically synthesized in small batches, but can be mass-produced. The dots can be distributed on a substrate by spin coating, either by hand or in an automated process. Large-scale production could use spray-on or roll-printing systems, dramatically reducing module construction costs.

Production

Early examples used costly molecular beam epitaxy processes. However, the lattice mismatch results in accumulation of strain and thus generation of defects, restricting the number of stacked layers. Droplet epitaxy growth technique shows its advantages on the fabrication of strain-free QDs. Alternatively, less expensive fabrication methods were later developed. These use wet chemistry (for CQD) and subsequent solution processing. Concentrated nanoparticle solutions are stabilized by long hydrocarbon ligands that keep the nanocrystals suspended in solution.

To create a solid, these solutions are cast down and the long stabilizing ligands are replaced with short-chain crosslinkers. Chemically engineering the nanocrystal surface can better passivate the nanocrystals and reduce detrimental trap states that would curtail device performance by means of carrier recombination. This approach produces an efficiency of 7.0%.

A more recent study uses different ligands for different functions by tuning their relative band alignment to improve the performance to 8.6%. The cells were solution-processed in air at room-temperature and exhibited air-stability for more than 150 days without encapsulation.

In 2014 the use of iodide as a ligand that does not bond to oxygen was introduced. This maintains stable n- and p-type layers, boosting the absorption efficiency, which produced power conversion efficiency up to 8%.

Market Introduction

Commercial Providers

Although quantum dot solar cells have yet to be commercially viable on the mass scale, several small commercial providers have begun marketing quantum dot photovoltaic products. Investors and financial analysts have identified quantum dot photovoltaics as a key future technology for the solar industry.

- Quantum Materials Corp. (QMC) and subsidiary Solterra Renewable Technologies are developing and manufacturing quantum dots and nanomaterials for use in solar energy and lighting applications. With their patented continuous flow production process for perovskite quantum dots, QMC hopes to lower the cost of quantum dot solar cell production in addition to applying their nanomaterials to other emerging industries.
- QD Solar takes advantage of the tunable band gap of quantum dots to create multi-junction solar cells. By combining efficient silicon solar cells with infrared solar cells made from quantum dots, QD Solar aims to harvest more of the solar spectrum. QD Solar's inorganic quantum dots are processed with high-throughput and cost-effective technologies and are more light- and air- stable than polymeric nanomaterials.
- UbiQD is developing photovoltaic windows using quantum dots as fluorophores. They have designed a luminescent solar concentrator (LSC) using near-infrared quantum dots which are cheaper and less toxic than traditional alternatives. UbiQD hopes to provide semi-transparent windows that convert passive buildings into energy generation units, while simultaneously reducing the heat gain of the building.

Safety Concerns

Many heavy-metal quantum dot (lead/cadmium chalcogenides such as PbSe, CdSe) semiconductors can be cytotoxic and must be encapsulated in a stable polymer shell to prevent exposure. Non-toxic quantum dot materials such as $AgBiS_2$ nanocrystals have been explored due to their safety and abundance; exploration with solar cells based with these materials have demonstrated comparable conversion efficiencies and short-circuit current densities. UbiQD's $CuInSe_2$-X quantum dot material is another example of a non-toxic semiconductor compound.

References

- "Fula solceller kan bli minne blott – svensk startup ska ta fram en "osynlig" solcell". Breakit (in Swedish). Retrieved 2019-06-04
- Solar-cell, technology: britannica.com, Retrieved 16 May, 2019
- Wu, Jiang; Yu, Peng; Susha, Andrei S.; Sablon, Kimberly A.; Chen, Haiyuan; Zhou, Zhihua; Li, Handong; Ji, Haining; Niu, Xiaobin (2015-04-01). "Broadband efficiency enhancement in quantum dot solar cells coupled with multispiked plasmonic nanostars". Nano Energy. 13: 827–835. Doi:10.1016/j.nanoen.2015.02.012
- Thin-film-solar-cell, technology: britannica.com, Retrieved 17 June, 2019
- Agarwal, S.; Nair, P.R. (2014-06-01). Performance optimization for Perovskite based solar cells. Photovoltaic Specialist Conference (PVSC), 2014 IEEE 40th. Pp. 1515–1518. Doi:10.1109/PVSC.2014.6925202. ISBN 978-1-4799-4398-2
- Organic-photovoltaics-research, solar, eere: energy.gov, Retrieved 18 July, 2019
- Snaith, Henry J. (2013). "Perovskites: The Emergence of a New Era for Low-Cost, High-Efficiency Solar Cells". The Journal of Physical Chemistry Letters. 4 (21): 3623–3630. Doi:10.1021/jz4020162

Solar Energy Power Generation

CHAPTER 4

Solar energy can be used for power generation through the conversion of the radiation of the sun into electricity by using solar photovoltaic cells. Some of the components of solar energy power generation are solar panels, solar power batteries and solar inverters. All these components of solar energy power generation have been carefully analyzed in this chapter.

Solar energy generation is one of fastest growing and most promising renewable energy sources of power generation worldwide. Nowadays, the electrical energy becomes one of the basic needs in our daily life, which makes increasing demand for it.

As a major source of electrical power generation fossil fuels are depleting day by day and also its usage raises serious environmental concerns. These reasons force the development of new energy sources which are renewable and ecologically safe.

The renewable energy sources include wind, solar, water, biomass and geothermal energy sources. Out of which, solar energy has the greatest potential in the long term and is predicted to play a major role in coming years. It is the cheapest method of generating electricity compared with other energy sources.

What is Solar Power and why to Choose?

Solar Power.

Solar power is the conversion of sun radiation into electricity through the use of solar photovoltaic cells. This conversion takes place in the solar cell by photovoltaic effect. As said by many experts that the amount of solar energy reaching the earth is more than 10000 times the current energy consumption by man.

Also, the power created by solar is sufficient for one year for the entire planet, if we could convert the 100 percent of the solar energy into electricity in one hour.

There are two ways of converting sunlight into electricity. In one method, solar energy is used simply as a source of heat. This heat is further used to produce the steam, which drives the steam turbine. This method of power generation is called solar thermal power generation.

In the second method, solar energy is directly converted into electricity using PV (or solar) cells as mentioned above. The PV cell is made with silicon semiconductor material.

Some of the factors for choosing the solar power generation are listed below:

- Solar energy is available freely and conveniently in nature and it needs no mains supply.
- Solar generation plant can be installed in a few months while the conventional power plants take several years to build an electricity generation plant.
- Solar power is clean energy as it produces no air or water pollution. Also, there are no moving parts to create noise pollution. Unlike fossil fuels, no toxic emissions are released into the atmosphere during solar energy power generation.
- Solar power has less running cost that means once the capital investment is made, there is no need for continues purchase of fossil fuels as the solar energy is effectively free in nature.

How Solar Power Generated from PV Cell?

A PV cell (can be called as a solar cell) is a semiconductor device that converts the sunlight energy into electricity without going through any energy conversion steps.

This conversion takes place by photovoltaic effect and hence they are called Photovoltaic (PV) cells. It generates voltage and current at its terminals when sunlight incident on it.

PV Cell.

The way and the amount of power generated by a solar cell depend on the sunlight falling on it. This also includes some factors such as intensity of light, angle at which the light falls on it and area of the cell.

The more is the power generated, if higher is the light intensity. If the area of the cell is more, the power generated is also more. And the optimum power is generated by it when light falling is perpendicular to the front side of the cell.

The solar cells are made with silicon semiconductor material and is treated with phosphorous and boron to make a thin silicon wafer. The wafer layers are then aligned together to make the solar cells, once they are doped.

Irrespective of the technology and material used, every solar cell has two terminals (positive and negative terminals) so as to take the electric current from it. Typically, a solar cell consists of front contact at the top, PN junction in the middle and back contact at the bottom.

Basically, the sunlight consists of bundles of photons, where each photon has a finite amount of energy. To generate the electricity from a solar cell, these photons must be absorbed by it. The energy of the photon and also the band-gap energy of semiconductor material decide the absorption of a photon.

Here is the term Electron-volt (eV) which is the unit of energy that expresses the photon energy and the band-gap energy of a semiconductor material.

The semiconductor material of the solar panel absorbs the photons in the sunlight. Due to this, electron-hole pairs are generated at the junction. When the solar cell is connected to the load, electrons and holes at the junction are separated from each other where the electrons are collected at the negative terminal and holes at positive terminal.

Thus the electric potential is built between the terminals and hence the voltage is developed across it. This further drives the current (DC) to the DC loads, inverter, or battery charging circuit.

If more photons are absorbed, greater will be the current generated. However, much of the solar radiation fall on the solar cell is not converted into electricity.

This is because light is composed of photons of different wavelengths. Some photons hit the solar cell and then reflected and prevented from entering to the cell. In some materials, generated electrons recombine with other molecules before being drawn into current.

Likewise, there are many reasons for the low conversion rate or efficiency. The conversion efficiency of the solar panels used in individual residences ranges from 6 to 10%.

And for large-scale installations and solar power plants, solar panels are designed with best material and technologies to achieve higher conversion efficiency ranging from 40 to 60%, but they are costliest.

A single solar cell of 4 cm^2 develops a voltage of 0.5 to 1 V and it can produce 0.7 W power when exposed to the sunlight. Typically, the best designed solar panel has a maximum efficiency of 25%.

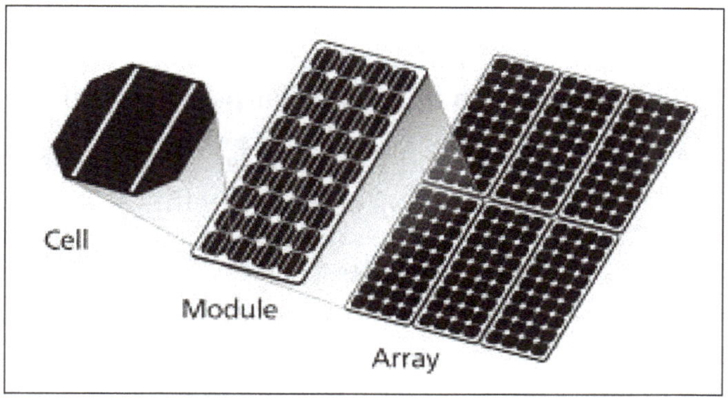

In order to generate high potential difference or voltage and more electric power, these individual cells are connected together that means some cells are connected in series and some are in parallel.

PV modules are formed by connecting the number of number of solar cells together. And several PV modules are connected together to make a PV array which can be used for small power as well as high power generation applications.

COMPONENTS OF SOLAR ENERGY ELECTRICITY GENERATION

The major components in solar energy electricity generation include:

Solar Panels

Solar cell modules or solar panels convert the solar energy into electricity. These are mounted in such a way that they collect maximum energy from the sun. Most solar panels are rated to a voltage 12V (a half volt PV cells are connected in series, inside of the solar panel to produce the high voltage say 12V).

Solar Panels.

The panels are connected in series to form a solar array that produces higher voltage, typically of 24 or 48V in standalone systems, or it can be several hundreds of volts in grid-connected systems.

If the panels are connected in parallel, the current delivered to the load will be more and hence the more power while maintaining the same voltage. Irrespective of series or parallel connection, the power rating of the system increases when multiple solar panels are connected together.

Here, a solar array is made with four 12V and 12 watt solar panels, where each panel produces a current of 1 A. Then this array would be rated as 48V, 48W with 1A current.

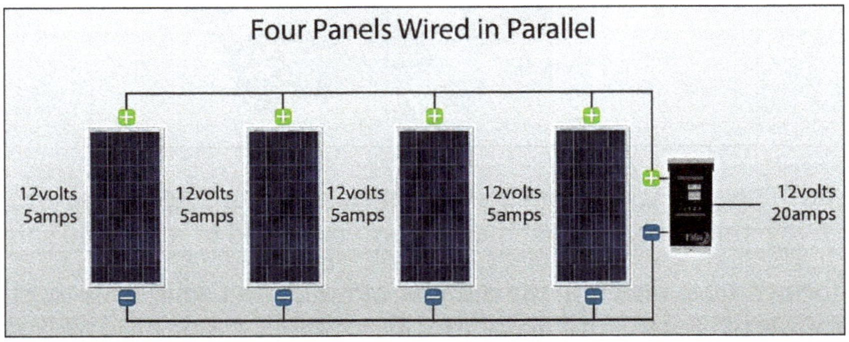

If the same rated solar panels are connected in parallel to form an array of four solar panels, then the solar array would be rated as 12V, 48W with 4 A current.

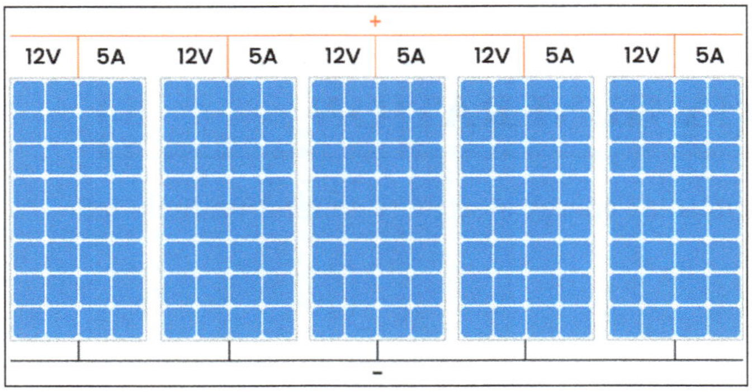

Batteries

Except the grid-connected system, all other solar energy power generation systems use batteries to store the energy generated from solar panels. Since the amount of solar power generated depends on the strength of the sunlight, batteries provide a constant source of power supply once it is fully charged.

Batteries.

Mostly, lead acid batteries are used in solar electric systems. Like solar panels, batteries can be connected together to form a battery bank.

These can be connected either series or parallel as similar to solar panels to achieve desired voltage, current and power ratings. The type of battery chosen depends on the energy requirements of a system and its budget.

Controller

It regulates the flow of current into and out of the battery. If the generated current overcharges the battery, it leads to damage in the battery. Moreover, if the battery is completely discharged, it will destroy the battery. Hence the solar controller prevents the batteries to undergo these conditions.

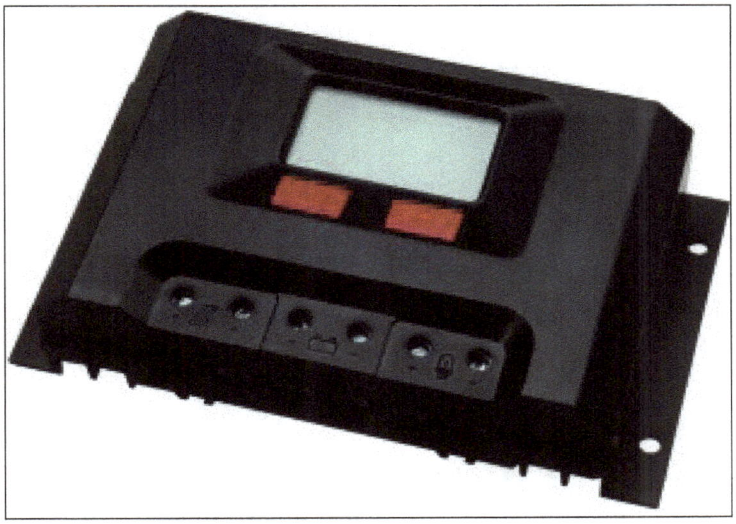

Solar Charge Controller.

The charge controller module balances the amount of electricity used to power appliance and lights with generated power. Also, it prevents the damage to the batteries due to overcharging and deep discharging. In addition, it gives the alarm when the module not functioning properly.

Inverter

The electricity generated from the solar panels is a Direct Current (DC), whereas the most electrical appliances work on Alternating Current (AC) and hence a converter is needed to convert DC to AC, nothing but an inverter.

Also, if the solar system is connected to the grid, the generated DC voltage must be converted into AC. So the inverter equipment converts the DC voltage to the AC and to the same voltage as that of grid or appliance rating.

As a recent invention, most individual solar panels are connected with micro inverters that provide a high AC voltage. These are suitable only for grid-connected systems and not suitable with battery backup systems.

Different Types of Solor Electric System

There are different configurations of solar electric system and are discussed below:

Off-grid/Standalone Solar Electric Systems

These are the most popular type of solar installations which are primarily designed to supplement or replace the conventional mains supply.

These are mainly used in the locations where there is no other source to provide power supply and hence these are used in remote locations and rural areas where it is difficult to get the power from grid extensions.

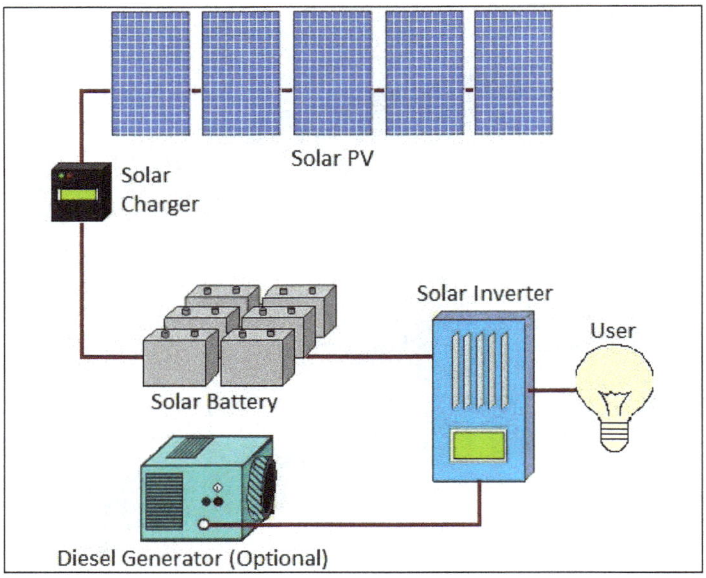

Generally, off-grid systems use solar power to charge the batteries, and this charge is then supplied to the load when needed. The battery power either directly operates the DC loads (DC lamps) or drives the power inverter that converts the DC power to AC power to operate the appliance like pumps, lighting equipment, refrigerators, etc.

This method is followed for any standalone system whether it is a pocket calculator or a complete off-grid home. Standalone systems are comparatively small and simple systems.

Grid-connected Solar Electric Systems

These systems effectively create a micro power station and are connected directly into the electricity grid. These are normally found in urban areas where power is readily available.

During the day it feeds the excess electricity generated into the grid and during evening and night it imports the electricity from the grid. Here the grid acts like a storage medium in which power is taken from the grid when needed.

It is to be noted that, grid-connected system doesn't have to supply enough electricity to cover entire power demand. So this system can be small or large depends on the owner's choice.

This system receives the payment for each kilowatt of power which is supplied to the electricity providers. This type of installation reduces the dependence on electric utilities and hence reduces the electricity bills.

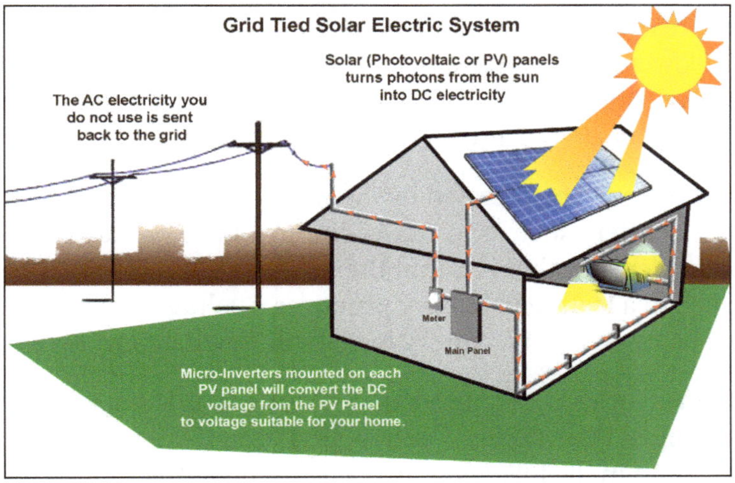

Grid-connected Solar Electric Systems.

The major components in this solar installation include a PV array or solar cells, inverter and the metering system. The main disadvantage of the grid-tied solar system is that it will switch off in the event of power cut, because this system is the part of utility grid and hence if there is a power cut, the power from the solar array is also switch off.

If this system won't stop, current flow back into the grid could lead to serious faults.

Central Grid-connected Solar Electric Systems

In a conventional grid system a variety of power sources such as gas, coal, water, etc. combined and it supplies the power to end user via transmission and distribution lines.

Likewise, a central grid connected system is directly connected to the transmission lines. These systems can be small (as 50kWp), large (as 100kWp) or somewhat higher range say 1GWp which all are directly connected to the transmission systems. Mostly, these are called as solar power plants.

Grid Fallback Solar Electric Systems

It combines the grid connected system with a bank of batteries. In this solar array generates the electricity, which in turn charges a battery bank. The battery power is then running the inverter which drives the loads through an inverter.

When the power in the batteries is not enough to drive the loads, the system automatically switches back to the grid power supply. Again, it switches back to battery power once the solar array recharges the batteries.

This system doesn't sell any power to any electrical utilities and the overall power generated used for domestic or residential systems alone.

Solar Power Battery

Batteries are often used in PV systems for the purpose of storing energy produced by the PV array during the day, and to supply it to electrical loads as needed (during the night and periods of cloudy weather). Other reasons because batteries are used in PV systems are to operate the PV array near its maximum power point, to power electrical loads at stable voltages, and to supply surge currents to electrical loads and inverters. In most cases, a battery charge controller is used in these systems to protect the battery from overcharge and overdischarge.

In stand-alone systems, the power generated by the solar panels is usually used to charge a lead-acid battery. Other types of battery such as nickel-cadmium batteries may be used, but the advantages of the lead-acid battery ensure that it is still the most popular choice. A battery is composed of individual cells; each cell in a lead-acid battery produces a voltage of about 2 Volts DC, so a 12 Volt battery needs 6 cells. The capacity of a battery is measured in Ampere-hours or Amp-hours (Ah).

To properly select batteries for use in stand-alone PV systems, it is important that system designers have a good understanding of their design features, performance characteristics and operational requirements. The information in the following sections is intended as a review of basic battery characteristics and terminology as is commonly used in the design and application of batteries in PV systems.

Battery Design and Construction

Battery manufacturing is an intensive, heavy industrial process involving the use of hazardous and toxic materials. Batteries are generally mass produced, combining several sequential and parallel processes to construct a complete battery unit. After production, initial charge and discharge cycles are conducted on batteries before they are shipped to distributors and consumers.

Manufacturers have variations in the details of their battery construction, but some common construction features can be described for most all batteries. Some important components of battery construction are described below:

Cell: The cell is the basic electrochemical unit in a battery, consisting of a set of positive and negative plates divided by separators, immersed in an electrolyte solution and enclosed in a case.

1. Active Material: The active materials in a battery are the raw composition materials that form the positive and negative plates, and are reactants in the electrochemical cell. The amount of active material in a battery is proportional to the capacity a battery can deliver. In lead-acid batteries, the active materials are lead dioxide (PbO_2) in the positive plates and metallic sponge lead (Pb) in the negative plates, which react with a sulfuric acid (H_2SO_4) solution during battery operation.

2. Electrolyte: The electrolyte is a conducting medium which allows the flow of current through ionic transfer, or the transfer of electrons between the plates in a battery. In a lead-acid battery, the electrolyte is a diluted sulfuric acid solution, either in liquid (flooded) form, gelled or absorbed in glass mats. In flooded nickel- cadmium cells, the electrolyte is an alkaline solution of potassium hydroxide and water. In most flooded battery types, periodic water additions are required to replenish the electrolyte lost through gassing. When adding water to batteries, it is very important to use distilled or de-mineralized water, as even the impurities in normal tap water can poison the battery and result in premature failure.

3. Grid: In a lead-acid battery, the grid is typically a lead alloy framework that supports the active material on a battery plate, and which also conducts current. Alloying elements such as antimony and calcium are often used to strengthen the lead grids, and have characteristic effects on battery performance such as cycle performance and gassing. Some grids are made by expanding a thin lead alloy sheet into a flat plate web, while others are made of long spines of lead with the active material plated around them forming tubes, or what are referred to as tubular plates.

4. Plate: A plate is a basic battery component, consisting of a grid and active material, sometimes called an electrode. There are generally a number of positive and negative plates in each battery cell, typically connected in parallel at a bus bar or inter-cell connector at the top of the plates. A pasted plate is manufactured by applying a mixture of lead oxide, sulfuric acid, fibers and water on to the grid. The thickness of the grid and plate affect the deep cycle performance of a battery. In automotive starting or SLI type batteries, many thin plates are used per cell. This results in maximum surface area for delivering high currents, but not much thickness and mechanical durability for deep and prolonged discharges. Thick plates are used for deep cycling applications such as for forklifts, golf carts and other electric vehicles. The thick plates permit deep discharges over long periods, while maintaining good adhesion of the active material to the grid, resulting in longer life.

5. Separator: A separator is a porous, insulating divider between the positive and negative plates in a battery, used to keep the plates from coming into electrical contact and short-circuiting, and which also allows the flow of electrolyte and ions between the positive and negative plates. Separators are made from micro-porous rubber, plastic or glass-wool mats. In some cases, the separators may be like an envelope, enclosing the entire plate and preventing shed materials from creating short circuits at the bottom of the plates.

6. Element: An element is defined as a stack of positive and negative plate groups and separators, assembled together with plate straps interconnecting the positive and negative plates.

7. Terminal Posts: Terminal posts are the external positive and negative electrical connections to a battery. A battery is connected in a PV system and to electrical loads at the terminal posts. In a lead-acid battery the posts are generally lead or a lead alloy, or possibly stainless steel or copper-plated steel for greater corrosion resistance. Battery terminals may require periodic cleaning, particularly for flooded designs. It is also recommended that the clamps or connections to battery terminals be secured occasionally as they may loosen over time.

8. Cell Vents: During battery charging, gasses are produced within a battery that may be vented to the atmosphere. In flooded designs, the loss of electrolyte through gas escape from the cell vents it a normal occurrence, and requires the periodic addition of water to maintain proper electrolyte levels. In sealed, or valve-regulated batteries, the vents are designed with a pressure relief mechanism, remaining closed under normal conditions, but opening during higher than normal battery pressures, often the result of overcharging or high temperature operation. Each cell of a complete battery unit has some type of cell vent.

9. Case: Commonly made from a hard rubber or plastic, the case contains the plates, separators and electrolyte in a battery. The case is typically enclosed, with the exception of inter-cell connectors which attach the plate assembly from one cell to the next, terminal posts, and vents or caps

which allow gassing products to escape and to permit water additions if required. Clear battery cases or containers allow for easy monitoring of electrolyte levels and battery plate condition. For very large or tall batteries, plastic cases are often supported with an external metal or rigid plastic casing.

Battery Types and Classifications

Many types and classifications of batteries are manufactured today, each with specific design and performance characteristics suited for particular applications. Each battery type or design has its individual strengths and weaknesses. In PV systems, lead-acid batteries are most common as we said before, due to their wide availability in many sizes, low cost and well understood performance characteristics. In a few critical, low temperature applications nickel-cadmium cells are used, but their high initial cost limits their use in most PV systems. There is no "perfect battery" and it is the task of the PV system designer to decide which battery type is most appropriate for each application. In general, electrical storage batteries can be divided into to major categories, primary and secondary batteries.

Primary Batteries

Primary batteries can store and deliver electrical energy, but cannot be recharged. Typical carbon-zinc and lithium batteries commonly used in consumer electronic devices are primary batteries. Primary batteries are not used in PV systems because they cannot be recharged.

Secondary Batteries

A secondary battery can store and deliver electrical energy, and can be recharged by passing a current through it in an opposite direction to the discharge current. Common leadacid batteries used in automobiles and PV systems are secondary batteries.

Lead-acid Battery Classifications

Many types of lead-acid batteries are used in PV systems, each having specific design and performance characteristics. While there are many variations in the design and performance of lead-acid cells, they are often classified in terms of one of the following three categories:

SLI Batteries

Starting, lighting and ignition (SLI) batteries are a type of lead-acid battery designed primarily for shallow cycle service, most often used to power automobile starters. These batteries have a number of thin positive and negative plates per cell, designed to increase the total plate active surface area. The large number of plates per cell allows the battery to deliver high discharge currents for short periods. While they are not designed for long life under deep cycle service, SLI batteries are sometimes used for PV systems in developing Batteries in PV Systems 29 countries where they are the only type of battery locally manufactured. Although not recommended for most PV applications, SLI batteries may provide up to two years of useful service in small stand-alone PV systems where the average daily depth of discharge is limited to 10-20%, and the maximum allowable depth of discharge is limited to 40-60%.

Motive Power or Traction Batteries

Motive power or traction batteries are a type of lead acid battery designed for deep discharge cycle service, typically used in electrically operated vehicles and equipment such as golf carts, fork lifts and floor sweepers. These batteries have a fewer number of plates per cell than SLI batteries, however the plates are much thicker and constructed more durably. High content lead-antimony grids are primarily used in motive power batteries to enhance deep cycle performance. Traction or motive power batteries are very popular for use in PV systems due to their deep cycle capability, long life and durability of design.

Stationary Batteries

Stationary batteries are commonly used in un-interruptible power supplies (UPS) to provide backup power to computers, telephone equipment and other critical loads or devices. Stationary batteries may have characteristics similar to both SLI and motive power batteries, but are generally designed for occasional deep discharge, limited cycle service.

There are several types of lead-acid batteries manufactured. The following sections describe the types of lead-acid batteries commonly used in PV systems.

Lead-antimony Batteries

Lead-antimony batteries are a type of lead-acid battery which use antimony (Sb) as the primary alloying element with lead in the plate grids. The use of lead-antimony alloys in the grids has both advantages and disadvantages. Advantages include providing greater mechanical strength than pure lead grids, and excellent deep discharge and high discharge rate performance. Lead-antimony grids also limit the shedding of active material and have better lifetime than lead-calcium batteries when operated at higher temperatures.

Disadvantages of lead-antimony batteries are a high self-discharge rate, and as the result of necessary overcharge, require frequent water additions depending on the temperature and amount of overcharge.

Most lead-antimony batteries are flooded, open vent types with removable caps to permit water additions.

They are well suited to application in PV systems due to their deep cycle capability and ability to take abuse, however they do require periodic water additions. The frequency of water additions can be minimized by the use of catalytic recombination caps or battery designs with excess electrolyte reservoirs. The health of flooded, open vent lead-antimony batteries can be easily checked by measuring the specific gravity of the electrolyte with a hydrometer.

Lead-antimony batteries with thick plates and robust design are generally classified as motive power or traction type batteries, are widely available and are typically used in electrically operated vehicles where deep cycle long-life performance is required.

Lead-calcium Batteries

Lead-calcium batteries are a type of lead-acid battery which use calcium (Ca) as the primary alloying element with lead in the plate grids. Like lead-antimony, the use of lead-calcium alloys in the grids has both advantages and disadvantages. Advantages include providing greater mechanical strength than pure lead grids, a low self-discharge rate, and reduced gassing resulting in lower water loss and lower maintenance requirements than for lead-antimony batteries. Disadvantages of lead-calcium batteries include poor charge acceptance after deep discharges and shortened battery life at higher operating temperatures and if discharged to greater than 25% depth of discharge repeatedly.

Flooded Lead-calcium: Open Vent

Often classified as stationary batteries, these batteries are typically supplied as individual 2 volt cells in capacity ranges up to and over 1000 ampere-hours. Flooded lead-calcium batteries have the advantages of low self-discharge and low water loss, and may last as long as 20 years in stand-by or float service. In PV applications, these batteries usually experience short lifetimes due to sulfation and stratification of the electrolyte unless they are charged properly.

Flooded Lead-calcium: Sealed Vent

Primarily developed as 'maintenance free' automotive starting batteries, the capacity for these batteries is typically in the range of 50 to 120 ampere-hours, in a nominal 12 volt unit. Like all lead-calcium designs, they are intolerant of overcharging, high operating temperatures and deep discharge cycles. They are "maintenance free" in the sense that you do not add water, but they are also limited by the fact that you cannot add water which generally limits their useful life. This battery design incorporates sufficient reserve electrolyte to operate over its typical service life without water additions. These batteries are often employed in small stand-alone PV systems such as in rural homes and lighting systems, but must be carefully charged to achieve maximum performance and life. While they are low cost, they are really designed for shallow cycling, and will generally have a short life in most PV applications.

An example of this type of battery that is widely produced throughout the world is the Delco 2000. It is relatively low cost and suitable for unsophisticated users that might not properly maintain their battery water level. However, it is really a modified SLI battery, with many thin plates, and will only provide a couple years of useful service in most PV systems.

Lead-antimony/Lead-calcium Hybrid

These are typically flooded batteries, with capacity ratings of over 200 ampere-hours. A common design for this battery type uses lead-calcium tubular positive electrodes and pasted lead-antimony negative plates.

This design combines the advantages of both lead-calcium and lead-antimony design, including good deep cycle performance, low water loss and long life. Stratification and sulfation can also be a problem with these batteries, and must be treated accordingly. These batteries are sometimes used in PV systems with larger capacity and deep cycle requirements. A common hybrid battery using tubular plates is the Exide Solar battery line manufactured in the United States.

Captive Electrolyte Lead-acid Batteries

Captive electrolyte batteries are another type of lead-acid battery, and as the name implies, the electrolyte is immobilized in some manner and the battery is sealed under normal operating conditions. Under excessive overcharge, the normally sealed vents open under gas pressure. Often captive electrolyte batteries are referred to as valve regulated lead acid (VRLA) batteries, noting the pressure regulating mechanisms on the cell vents. Electrolyte cannot be replenished in these battery designs, therefore they are intolerant of excessive overcharge.

Captive electrolyte lead-acid batteries are popular for PV applications because they are spill proof and easily transported, and they require no water additions making them ideal for remote applications were maintenance is infrequent or unavailable. However, a common failure mode for these batteries in PV systems is excessive overcharge and loss of electrolyte, which is accelerated in warm climates. For this reason, it is essential that the battery charge controller regulation set points are adjusted properly to prevent overcharging.

This battery technology is very sensitive to charging methods, regulation voltage and temperature extremes.

Optimal charge regulation voltages for captive electrolyte batteries varies between designs, so it is necessary to follow manufacturers recommendations when available. When no information is available, the charge regulation voltage should be limited to no more than 14.2 volts at 25 C for nominal 12 volt batteries.

A benefit of captive or immobilized electrolyte designs is that they are less susceptible to freezing compared to flooded batteries. Typically, lead-calcium grids are used in captive electrolyte batteries to minimize gassing, however some designs use lead-antimony/calcium hybrid grids to gain some of the favorable advantages of lead-antimony batteries.

In the United States, about one half of the small remote PV systems being installed use captive electrolyte, or sealed batteries. The two most common captive electrolyte batteries are the gelled electrolyte and absorbed glass mat designs.

Gelled Batteries

Initially designed for electronic instruments and consumer devices, gelled lead-acid batteries typically use lead-calcium grids. The electrolyte is 'gelled' by the addition of silicon dioxide to the electrolyte, which is then added to the battery in a warm liquid form and gels as it cools. Gelled batteries use an internal recombinant process to limit gas escape from the battery, reducing water loss. Cracks and voids develop within the gelled electrolyte during the first few cycles, providing paths for gas transport between the positive and negative plates, facilitating the recombinant process.

Some gelled batteries have a small amount of phosphoric acid added to the electrolyte to improve the deep discharge cycle performance of the battery. The phosphoric acid is similar to the common commercial corrosion inhibitors and metal preservers, and minimizes grid oxidation at low states of charge.

Absorbed Glass Mat (AGM) Batteries

Another sealed, or valve regulated lead-acid battery, the electrolyte in an AGM battery is absorbed in glass mats which are sandwiched in layers between the plates. However, the electrolyte is not gelled. Similar in other respects to gelled batteries, AGM batteries are also intolerant to overcharge and high operating temperatures. Recommended charge regulation methods stated above for gelled batteries also apply to AGMs.

A key feature of AGM batteries is the phenomenon of internal gas recombination. As a charging lead-acid battery nears full state of charge, hydrogen and oxygen gasses are produced by the reactions at the negative and positive plates, respectively. In a flooded battery, these gasses escape from the battery through the vents, thus requiring periodic water additions. In an AGM battery the excellent ion transport properties of the liquid electrolyte held suspended in the glass mats, the oxygen molecules can migrate from the positive plate and recombine with the slowly evolving hydrogen at the negative plate and form water again. Under conditions of controlled charging, the pressure relief vents in AGM batteries are designed to remain closed, preventing the release of any gasses and water loss.

Calcium-Calcium

The next stage has dominated on a wide front in recent years and is called Calcium-Calcium. This involves the antimony on both the negative and positive plates being replaced by calcium alloy. The benefits are obvious. The fluid loss of the battery is about 80 % lower than that of antimony batteries and the self-discharge is lower, i.e. they can remain unused for longer periods without losing a lot of their charge. The disadvantage is that they are more demanding when charging if they have been over-discharged. Efforts made in avoiding the gassing had a positive affect, namely that the bubbles moved about in the acid causing it to be thoroughly mixed when charging. Without these bubbles the acid can stratify at different densities, acid weights, and is quite a common phenomenon. An acid weight of 1.35 or more at the bottom and maybe 1.17 at the top when you are looking for an even weight of 1.28 may cause the battery to be affected by sulphating and increased grid corrosion despite the battery being apparently fully charged.

Lead-acid Battery Chemistry

Now that the basic components of a battery have been described, the overall electrochemical operation of a battery can be discussed. The basic lead-acid battery cell consists of sets positive and negative plates, divided by separators, and immersed in a case with an electrolyte solution. In a fully charged lead-acid cell, the positive plates are lead dioxide (PbO_2), the negative plates are sponge lead (Pb), and the electrolyte is a diluted sulfuric acid solution. When a battery is connected to an electrical load, current flows from the battery as the active materials are converted to lead sulfate ($PbSO_4$).

Formation

Forming is the process of initial battery charging during manufacture. Formation of a lead-acid battery changes the lead oxide (PbO) on the positive plate grids to lead dioxide (PbO_2), and to metallic sponge lead (Pb) on the negative plates. The extent to which a battery has been formed during manufacture dictates the need for additional cycles in the field to achieve rated capacity.

Specific Gravity

Specific gravity is defined as the ratio of the density of a solution to the density of water, typically measured with a hydrometer. By definition, water has a specific gravity of one. In a lead-acid battery, the electrolyte is a diluted solution of sulfuric acid and water. In a fully charged battery, the electrolyte is approximately 36% sulfuric acid by weight, or 25% by volume, with the remainder water. The specific gravity of the electrolyte is related to the battery state of charge, depending on the design electrolyte concentration and temperature.

In a fully charged flooded lead-acid battery, the specific gravity of the electrolyte is typically in the range of 1.250 to 1.280 at a temperature of 27 °C, meaning that the density of the electrolyte is between 1.25 and 1.28 times that of pure water. When the battery is discharged, the hydrogen (H^+) and sulfate (SO_4^{2-}) ions from the sulfuric acid solution combine with the active materials in the positive and negative plates to form lead sulfate ($PbSO_4$), decreasing the specific gravity of the electrolyte. As the battery is discharged to greater depths, the sulfuric acid solution becomes diluted until there are no ions left in solution. At this point the battery is fully discharged, and the electrolyte is essentially water with a specific gravity of one.

Concentrated sulfuric acid has a very low freezing point (less than -50 °C) while water has a much higher freezing point of 0 °C. This has important implications in that the freezing point of the electrolyte in a lead- acid battery varies with the concentration or specific gravity of the electrolyte. As the battery becomes discharged, the specific gravity decreases resulting in a higher freezing point for the electrolyte.

Lead-acid batteries used in PV systems may be susceptible to freezing in some applications, particularly during cold winters when the batteries may not be fully charged during below average insolation periods.

The PV system designer must carefully consider the temperature extremes of the application along with the anticipated battery state of charge during the winter months to ensure that lead-acid batteries are not subjected to freezing.

Adjustments to Specific Gravity

In very cold or tropical climates, the specific gravity of the sulfuric acid solution in lead-acid batteries is often adjusted from the typical range of 1.250 to 1.280. In tropical climates where freezing temperatures do not occur, the electrolyte specific gravity may be reduced to between 1.210 and 1.230 in some battery designs.

This lower concentration electrolyte will lessen the degradation of the separators and grids and prolong the battery's useful service life. However, the lower specific gravity decreases the storage capacity and high discharge rate performance of the battery. Generally, these factors are offset by the fact that the battery is generally operating at higher than normal temperatures in tropical climates.

In very cold climates, the specific gravity of the electrolyte may be increased above the typical range of 1.250 to 1.280 to values between 1.290 and 1.300. By increasing the electrolyte concentration, the electrochemical activity in the battery is accelerated, improving the low temperature capacity and lowers the potential for battery freezing. However, these higher specific gravities generally reduce the useful service life of a battery.

While the specific gravity can also be used to estimate the state of charge of a lead-acid battery, low or inconsistent specific gravity reading between series connected cells in a battery may indicate sulfation, stratification, or lack of equalization between cells. In some cases a cell with low specific gravity may indicate a cell failure or internal short-circuit within the battery. Measurement of specific gravity can be a valuable aid in the routine maintenance and diagnostics of battery problems in stand-alone PV systems.

Nickel-cadmium Batteries

Nickel-cadmium (Ni-Cad) batteries are secondary, or rechargeable batteries, and have several advantages over lead-acid batteries that make them attractive for use in stand-alone PV systems. These advantages include long life, low maintenance, survivability from excessive discharges, excellent low temperature capacity retention, and non-critical voltage regulation requirements. The main disadvantages of nickel- cadmium batteries are their high cost and limited availability compared to lead-acid designs.

A typical nickel-cadmium cell consists of positive electrodes made from nickel-hydroxide (NiO(OH))and negative electrodes made from cadmium (Cd) and immersed in an alkaline potassium hydroxide (KOH) electrolyte solution. When a nickel-cadmium cell is discharged, the nickel hydroxide changes form ($Ni(OH)_2$) and the cadmium becomes cadmium hydroxide ($Cd(OH)_2$). The concentration of the electrolyte does not change during the reaction so the freezing point stays very low.

Nickel-cadmium Battery Chemistry

Following are the electrochemical reactions for the flooded nickel-cadmium cell:

At the positive plate or electrode:

$$2NiO(OH)+2H_2O+2e^- \Leftrightarrow 2Ni(OH)_2+2OH^-$$

At the negative plate or electrode:

$$Cd+2OH^- \Leftrightarrow Cd(OH)_2+2e^-$$

Overall nickel cadmium cell reaction:

$$Cd+2NiO(OH)+2H_2O \Leftrightarrow Cd(OH)_2+2Ni(OH)_2$$

Notice these reactions are reversible and that the elements and charge are balanced on both sides of the equations. The discharge reactions occur from left to right, while the charge reactions are reversed.

The nominal voltage for a nickel-cadmium cell is 1.2 volts, compared to about 2.1 volts for a lead-acid cell, requiring 10 nickel-cadmium cells to be configured in series for a nominal 12 volt battery. The voltage of a nickel-cadmium cell remains relatively stable until the cell is almost completely discharged, where the voltage drops off dramatically. Nickel-cadmium batteries are commonly subdivided in to two primary types; sintered plate and pocket plate.

Ni-Cd Types

Sintered Plate Ni-Cads

Sintered plate nickel cadmium batteries are commonly used in electrical test equipment and consumer electronic devices. The batteries are designed by heat processing the active materials and rolling them into metallic case. The electrolyte in sintered plate nickel-cadmium batteries is immobilized, preventing leakage, allowing any orientation for installation. The main disadvantage of sintered plate designs is the so called 'memory effect', in which a battery that is repeatedly discharged to only a percentage of its rated capacity will eventually 'memorize' this cycle pattern, and will limit further discharge resulting in loss of capacity. In some cases, the 'memory effect' can be erased by conducting special charge and discharge cycles, regaining some of its initial rated capacity.

Pocket Plate Ni-Cads

Large nickel cadmium batteries used in remote telecommunications systems and other commercial applications are typically of a flooded design, called flooded pocket plate. Similar to flooded lead-acid designs, these batteries require periodic water additions, however, the electrolyte is an alkaline solution of potassium hydroxide, instead of a sulfuric acid solution. Advantages:These batteries can withstand deep discharges and temperature extremes much better than lead-acid batteries, and they do not experience the 'memory effect' associated with sintered plate Ni-Cads. The main disadvantage of pocket plate nickel cadmium batteries is their high initial cost, however their long lifetimes can result in the lowest life cycle cost battery for some PV applications.

Main Degradation Mechanisms of Solar Batteries

In stand-alone PV systems special attention must be paid to the battery bank, which is often said to be the weakest component of the system and the highest contributor to its life cycle cost. For this reason, designers, project managers are very concerned by the evaluation of the battery lifetime.

The main objective of a cycling test procedure is to evaluate the battery lifetime, expressed in terms of reduced capacity, by reproducing a typical use of the battery in the field. The test procedures induce ageing of the battery by accelerating the degradation (sulphation, corrosion, stratification and softening) of the battery grid and the active mass. The analysis of the results shows that most of the cycling procedures lead to significant battery sulphation, but almost none of them accelerate battery corrosion. In fact, there is no test procedure where the four degradation mechanisms lead to a significantly high level of degradation. It is therefore advisable for project managers to select several methods in order to ascertain the most likely battery degradation pattern for a given application and then be able to choose the right battery.

In terms of battery technology, there is a wide range of efficiency and lifetime values for various batteries of the same technology. This is due to differences in the type of grid alloy, the active mass composition, the manufacturing process or the cell geometry. In addition, the longer and better service of tubular batteries was confirmed by the analysis.The behaviour of tubular batteries is, in general, not very dependent on the selected test procedure, as this kind of battery is more sensitive to sulphation than to softening and corrosion. However, flat plate solar batteries are sensitive to all types of degradation and their cycling life is much more dependent on the test procedure.

Nevertheless there is still a lot of work to do on the topic of accelerated battery cycling life evaluation. There is a strong need for a deeper international collaboration to focus on a limited number of cycling procedures, gather more results on battery tests, search for other means to accelerate the tests (key parameters, modelling studies) and lead to the standardization of these cycling tests.

Four main degradation mechanisms can occur in solar battery in operating conditions:

- Softening of the active mass,
- Corrosion of the grid,
- Sulphation,
- Stratification of the electrolyte.

They are briefly described:

Softening

When a battery is submitted to successive cycles of charge/discharge, repeated volume variations of the active mass create some changes in its morphology such as a modification of the cohesion, the distribution of lead oxides (PbO_2) and the size of crystals. These changes lead to a loose of connection between the aggregates of lead dioxide and a softening of the active mass. The first consequence is a loss of capacity caused by a lower participation of this active mass in the electrochemical reactions. The extreme situation is obtained when there is no more connection between the aggregates and they fall to the bottom of the battery container. This process is called shedding.

Corrosion

When a lead-acid battery is in a high voltage condition (end of charge or overcharge), the oxygen produced at the positive plate leads to the formation of an oxide layer at the interface between the current collector, or grid, and the active mass. A similar corrosion layer is formed when the battery is left in open circuit conditions for a long time.

The corrosion layer is resistive which affects the current collection by the grid. There are many consequences such as:

- Decrease of electronic exchanges (the layer forms a barrier to ionic diffusion),
- Increase of the internal resistance,
- Decrease of the charge acceptance,
- Decrease of the battery capacity,
- Mechanical consequences,
- Grids becoming thin and fragile.

Sulphation

The "so-called" irreversible sulphation phenomenon is the presence of nonrechargeable lead sulphate crystals in positive and negative active masses. Lead sulphate is formed during the discharge phase of the battery. When the battery remains at low states of charge, a process of re-crystallisation of the lead sulphate crystals happens and affects their characteristics: $PbSO_4$ crystals at the positive and negative plates become bigger and are less connected with the active mass; thus they are no more rechargeable. This leads to a loss of capacity.

Stratification

The stratification of the electrolyte is the presence of a vertical gradient of the concentration of sulphuric acid due to the operation of the battery. Pure sulphuric acid formed during the charge phase has a higher density than the bulk electrolyte and tends to fall down to the bottom part of the battery. This phenomenon is favoured by deep discharges and recharges, but decreases during overcharge as the electrolyte is mixed by oxygen and hydrogen gas bubbles. The stratification also depends on the characteristics of the battery such as: the grid alloy (batteries with lead-antimony grids are less sensitive to stratification); the plate Grid Corrosion layergeometry (stratification is favoured with the height of the plates), and the compression level of the plates stack (separators slow the descent of acid to the bottom of the battery). The stratification of the electrolyte leads to a loss of capacity (bottom part of the battery less recharged) and to the forming of irreversible lead sulphate crystals. The stratification phenomenon does not occur, or only at a low level, in VRLA batteries because the electrolyte is immobilised in Absorptive Glass Material (AGM) separators or in a gel.

This photography is obtained through the use of a radioactive tracer that enables to display the distribution of irreversible lead sulphates at the interface active mass/electrolyte: the lead sulphate crystal density increases from "white" to "red", showing here a clear difference of concentration between the lower and upper parts of the plates. This methodology has been developed by Genec.

Battery System Design and Selection Criteria

Battery system design and selection criteria involves many decisions and tradeoffs. Choosing the right battery for a PV application depends on many factors. While no specific battery is appropriate for all PV applications, common sense and a careful review of the battery literature with respect to the particular application needs will help the designer take the decision.

Some decisions on battery selection may be easy to arrive at, such as physical properties, while other decisions will be much more difficult and may involve making tradeoffs between desirable and undesirable battery features. With the proper application of this knowledge, designers should be able to differentiate among battery types and gain some application experience with batteries they are familiar with.

Life Expectancy of Solar Rechargeable Batteries

Batteries used in solar power systems can last from one to twenty years, depending on type and how they are used. Factors such as improper charging and temperature all take their toll on solar batteries.

Type

Solar batteries are typically deep-cycle, which allows them to discharge up to 80% before recharging. Used properly, they can last several years. Each battery is rated for a certain number of full cycles, i.e. it can be discharged then recharged a certain number of times.

Typical Lifespan

According to Northern Arizona Wind & Sun, an AGM deep-cycle battery lasts four to seven years and a gelled deep cycle battery lasts two to five years. A regular deep cycle battery lasts four to eight years.

Industrial Batteries

Less common options are industrial batteries. Premium deep cycle batteries may last 7 to 15 years, industrial type batteries may last 10 to 20 years and surplus telephone (float) batteries can last 2 to 20 years.

Abuse

The lifespan of batteries can be greatly reduced by abuse. Lifespan is reduced by storing batteries in temperatures over 77 °C, because they discharge more quickly. Overcharging is also an easy way to kill batteries.

Efficiency

Battery life can be extended by fully discharging them before recharging, i.e using them as designed. Keeping the batteries at a steady temperature, neither overcharging nor excessively draining them, all help.

Battery Standards

Standards are produced to facilitate replacement of battery cells (standardised dimensions and designs) and the understanding of performance data (standardised testing methods), etc. The standardisation bodies represent groups of companies and organisations with a common interest in the production or availability of comparable products from multiple suppliers or comparable measures of different parameters.

IEC Standards

The International Electrotechnical Committee (IEC) develops and publishes standards for batteries. Several standards are covering the lead-acid battery and its different versions and applications. The most relevant standards are:

- IEC 95 Lead-acid starter batteries.
- IEC 254 Lead-acid traction batteries.
- IEC 896 Stationary lead-acid batteries.

- IEC 1429 Marking of secondary cells and batteries with the international recycling symbol.
- IEC 1431 Guide for the use of monitor systems for lead-acid traction batteries.

Other Standards

Other bodies are also developing standards for specific purposes or for geographical regions. It is important to investigate if some standards are valid for your application and put specific demands on the installation at its specific location.

Standards are for example produced by:

- CENELEC produces standards for Europe.
- SAE produces standards for the (American) car industry.
- DIN produces German national standards.
- JISC produces Japanese national standards.

Safety Precautions

The lead-acid battery is an energy pack. It is quite sensitive to abuse. When treated in a wrong way, it is easy to pass over the limits to unstable conditions. Never stress the battery mechanically, electrically or thermally beyond the limits.

Always use Insulated Tools

The tools used for connecting batteries have to be insulated to avoid the potential shorting of batteries. Plastic tubes or plastic tape can easily be used for home manufacturing of insulated tools.

Protect Eyes and Skin

- Safety goggles must always be worn when working with batteries.
- Rubber gloves should preferably be used to avoid skin contact with the electrolyte.
- Special clothing is recommended to avoid droplets of acid destroying the normal dress. Such droplets are not always recognised until the clothes are washed and the acid has already made holes in the dress.

Large amounts of water should be used for rinsing if acid has come into contact with the skin, clothes or the construction of the installation. It is very important to have a water bottle available within easy reach in the working area.

Protection against Explosion

Explosions may result from different abuse situations. An explosion forces the cell case to burst and parts will fly around together with a spray of acid. A fire may also start as a result of the explosion.

The battery should be protected against short circuit by an electrical fuse sized correctly to protect the battery from currents above the maximum load and charge values. The fuse is preferably placed in the middle of the voltage string in the battery pack. The location of the fuse in the middle of the voltage string will be safer if a shortcut only involves some part of the battery. In a small 12V monoblock battery system the fuse should be placed as close as possible to the positive terminal.

When assembling the parts of the installation, never connect the battery terminals until all cabling is finished and checked for faults.

Mechanical abuse may well result in an explosion. Protect the battery from pieces that could fall down on the battery or be injected from the side. Mechanical damage to the battery may result in an explosion, which starts as an internal short circuit when the plates or connectors of different polarity are contacting each other as a result of the damage.

A gas explosion is commonly created by an external spark. When batteries are fully charged and in the gassing mode, the internals of the cells are filled with an explosive gas mixture and gas is issuing through the vents. This gas is highly explosive and a cigarette, a match or even by an electrical spark, can easily ignite it. Such a spark may result from a static discharge when two materials are statically charged or from an electrical circuit.

A gas explosion in one cell will immediately propagate from one cell to all other cells, damaging the whole battery.

How to Carry a Battery

Now and then it is necessary to move a battery from one place to another. Since a battery is a heavy piece of equipment it is recommended to use a special strap which is placed around the terminals. Never try to carry a battery without it.

First of all this could do severe harm to your back and secondly there is a definitive risk of you dropping the battery on the floor. The result is possible harm to your feet not only from the impact when the battery reaches the floor but also from sulphuric acid if the case breaks.

Disposal of the Battery at end of Life

Batteries are considered hazardous items as they contain toxic materials such as lead, acids and plastics which can harm humans and the environment.

For this reason, it is important to reuse the materials for the manufacturing of new batteries.

Such processes for recovery of the metal content of lead-acid batteries are established. If batteries are thrown away, they may leak poisonous metals and acid into the soil.

Under no circumstances should a batteries be disposed of in landfills, or the electrolyte allowed to seep into the ground, or the battery burned.

Be careful to collect old batteries for metal recovery or renovation (mainly in developing countries). They may represent a significant value in some countries.

Environmental Impacts of Batteries

The main objective of this important section is to know more about the environmental impacts of batteries .We will see which one has the best environmental profile, we will try to find out if it is really applied on a big scale and what are pros and cons.

Ultimately, we will discuss the environmental impacts of them.

Eight battery technologies are evaluated: lithium-ion (Li-ion), sodium–sulphur (NaS), nickel–cadmium (NiCd), nickel–metal hydride (NiMH), lead–acid (PbA), vanadium-redox (VRB), zinc–bromine (ZnBr) and polysulfide-bromide (PSB).

The energy requirements for production and transport of PV-battery systems that use the different battery technologies differ by up to a factor of three. Production and transport of batteries contribute 24–70% to the energy requirements, and the PV array contributes 26–68%. The contribution from other system components is less than 10%. The contribution of transport to energy requirements is 1–9% for transportation by truck, but may be up to 73% for air transportation. The energy requirement for battery production and transport is dominant for systems based on NiCd, NiMH and PbA batteries.

The energy requirements for these systems are, therefore, sensitive to changes in battery service life and gravimetric energy density. For systems with batteries with relatively low energy requirement for production and transportation (Li-ion, NaS, VRB, ZnBr, PSB), the battery charge–discharge efficiency has a larger impact.

Today, lead–acid and nickel–cadmium batteries are commonly used in PV Systems.

Some emerging battery technologies may also be suitable for storage of renewable energy, such as different types of redox flow batteries and high temperature sodium–sulphur batteries(the NaS battery is commercially produced and is used in applications for power quality and uninterrupted power supply).

The PSB, VRB and ZnBr batteries are redox flow batteries based on liquid electrolytes, which are pumped into the battery stack. The size of the stack determines the power rating, and the volume of the electrolyte determines the storage capacity. The PSB battery technology has not yet been demonstrated in commercial operation, and the construction of two plants with a power rating of 12–15 MW and an energy capacity of 120 MWh was stopped in 2003 . The VRB battery technology has been demonstrated by different manufacturers for load levelling and PV applications with power ratings up to 1.5 MW and an energy capacity up to 5 MWh.

Demonstration units of the ZnBr battery have been built for PV applications with the ratings of 50 kW/100 kW h and 250 kW/500 kW h.

There are also included NiMH and Li-ion batteries since their use in small scale applications is growing rapidly. However, their relatively high production cost has precluded them from employment in applications for storage of several hundreds of kiloWatt hour.

Identification of the important parameters in PV applications can be used to direct research and product improvements, and comparison of different battery technologies can be used to guide battery choice for specific user conditions.

For energy technologies, the energy requirement for producing equipment is an important performance parameter. Large energy requirements in comparison to energy output will limit the range of possible applications to small niches. Energy requirements for producing PV modules have been studied and debated since the early 1970s, while batteries have gained less attention.

In a study of solar home systems, it was concluded that lead–acid batteries contribute significantly to the energy requirements.

The purpose of this study is to provide an energy analysis to enable comparison of different battery technologies in renewable energy applications. By quantifying energy efficiencies and the energy requirements for manufacturing the different systems, increased awareness may lead to improved energy management of energy storage systems.

The battery parameters investigated are battery charge–discharge efficiency, service life, gravimetric energy density and energy requirements for production and transport of the batteries.

The PV modules are assumed to be based on multi-crystalline silicon (mc-Si), i.e. the PV technology that currently dominates the market.

Corrections for power or temperature deviation, incomplete utilisation of irradiation etc. are not explicitly considered for these components but are assumed to be included in these efficiency ranges.

The charge–discharge efficiency is highest for the Li-ion battery and lowest for the PSB battery. For batteries requiring pumps and auxiliary components, these losses are included.

The annual solar irradiation (H) was assumed to be 1.7 MWh/m^2 yr, representing medium irradiation levels, which can be found in Southern Europe and large parts of the USA. The maximum solar irradiance (Sp) is assumed to be 1000 W/m^2.

Service Life of the Components

The end of battery service life is when the battery capacity has reached 80% of its initial capacity or when it fails to function. The effects of ambient temperature on the performance and service life of redox flow batteries and the NaS battery are limited since their operating temperatures are regulated by pumping of the electrolytes or by thermal management systems.

When Li-ion, NiCd, NiMH and PbA batteries are used in applications with shallow cycling, their service life normally will be limited by float life.

In systems where the cycling is deep, but occurs only a few times a year, temperature dependent corrosion processes is the normal life limiting factor, even for batteries with low cycle life.

In systems with deep daily cycling, the cycle life determines the service life of the battery. Considering life limiting conditions, the VRB, Li-ion and NaS have the highest service lives.

As the temperature increases, the electrochemical activity of the battery increases as well as the speed of the natural ageing of the active material. Accelerated life tests at elevated temperature, correlated with corrosion studies, provide a basis for estimating service life. Service temperature is the key factor in determining corrosion. As a rule, for PbA batteries every 10 °C increase in temperature reduces service life by 50%. Valve regulated lead-acid batteries have been shown to have

>15 years float service life at 20 °C and >10 years at 25 °C. The rate of ageing for NiCd batteries is about 20% reduction in life for 10 °C increase in temperature. NiCd is less affected than PbA since increased electrochemical activity has little effect on the steel structural components of the NiCd electrode assembly.

The PSB and VRB batteries are little affected by varying ambient temperatures, since pumps are used to circulate the electrolyte to heat exchangers, which maintain their operating temperatures.

No degradation of the NaS battery has been documented, but the temperature of the electronic control systems has to be controlled. Elevated temperature may degrade the plastic materials of the ZnBr battery, but no degradation has been quantified.

Energy Requirements for Production and Transport of the PV-battery System Components

The gravimetric density for the PV array, including module frame, was estimated at 9 kg/m² for roof integrated systems and 12 kg/m² for ground mounted systems. The PSB and VRB are the heaviest ones (24–49 tons) followed by PbA (15–24 tons). The mass of these systems is 5–8 times higher than that of the Li-ion battery, which has the lowest weight. The weight fraction of the electrolyte active material, y, is 0.32–0.40 for the redox flow batteries (PSB, VRB and ZnBr) and 0 for the other batteries.

The energy requirements for mc-Si modules are estimated at 4200 MJ/m² or 32 MJ/W.

The energy requirements for production of single crystalline silicon (sc-Si) and amorphous silicon (a-Si) modules are 16–20% higher and 0–17% lower, respectively, than that for mc-Si.

Module frames are assumed to use 2.5 kg Al/m² module corresponding to 500 MJ/m².

Energy requirements for array supports in roof integrated and ground mounted systems have been estimated to be 700 MJ/m² and 1800 MJ/m², respectively. The total energy requirements for production of PV arrays (module, frame and array support) were calculated to be 5400 MJ/m² (roof integrated) and 6500 MJ/m² (ground mounted), corresponding to 45–54 MJ/W.

Energy requirements for producing the inverter, charger and AC were estimated to be 1 MJ/ W, resulting in 70–100 MJ/kg.

Energy requirements for production of batteries were assessed from cradle to gate, including materials production and battery manufacturing. To enable a thorough evaluation of different LCA studies, it is needed to report the following parameters:

(1) battery mass, (2) battery capacity, (3)choice of system boundaries (geographical, temporal, technological), (4) electricity´s share of gross primary energy requirements, assumed efficiency of electricity production and method for adding up different energy qualities, (5) recycling rate of used batteries, (6) energy requirements for battery manufacturing processes and production of virgin and recycled materials, respectively, (8) battery design, (9) battery material composition and (10) the allocation principles for multi- output processes.

Based on published life cycle assessments and estimates, the energy requirements for production of different batteries were estimated. The energy requirements for battery production were scattered between studies, and there are uncertainties on the material requirements and manufacturing processes, particularly for the Li-ion, NaS and all the redox flow batteries.

The energy requirements of batteries with active materials in solid phase are assumed to be independent of the required PV-battery system output power by changing the cell configuration and voltage of the battery strings.

The energy requirements for production of batteries range from 0.87 MJ/Wh (PbA based on recycled materials) to 4.1 MJ/Wh (NiCd based on virgin materials). Differences are partly explained by the energy intensity of materials production. For the NiCd battery, steel and nickel contribute 60–70% to the energy requirements of materials production. Production of nickel requires 2–8 times more energy than that for the lead used in PbA batteries.

The energy requirements for manufacturing processes contribute 33–78% of the energy requirements for battery production, resulting in relatively small changes between virgin and recycled materials.

The energy requirements for battery manufacturing are assumed to be constant while the energy requirements for materials vary depending on the recycling of materials. The production of recycled materials requires 32–75% less energy than virgin materials since the energy for extraction from mines and reduction of metals are allocated to the first material life cycle. For redox flow batteries, it is assumed that the energy requirements for production of stack and other battery components are the same as for electrolyte production.

At the end of the service life of redox flow batteries, they can be renovated by renewing the stacks while the electrolyte with active material is assumed to be used indefinitely. Batteries based on active materials in solid phase (Li-ion, NaS, NiCd, NiMH, PbA) need to be manufactured and replaced completely.

Contributing Components

Production and transportation of batteries contributes 24–70% of the energy requirements of the PV-battery system. The relative contribution from the production of batteries is lowest for the ZnBr battery and highest for the NiMH battery.

The contribution of production and transport of the PV array is 26–68% (NiMH–ZnBr). The highest absolute energy requirement for PV array production is 88–96 GJ/yr for the redox flow batteries due to their relatively low efficiency, resulting in the need for a larger PV array and charge regulator. Production and transport of the charge regulator and inverter contribute 2–4%, respectively.

The contribution of transport of all the components to the gross energy requirement is low (0.9–8.9%) for 3000 km transport by heavy truck. The lowest energy requirement for transport is for the ZnBr battery due to its high energy density and the possibility of recycling the electrolyte. The transport of PbA batteries contributes 8.9% to the gross energy requirement since these batteries have a relatively low energy density and cycle life, and therefore a larger mass of batteries has to be transported.

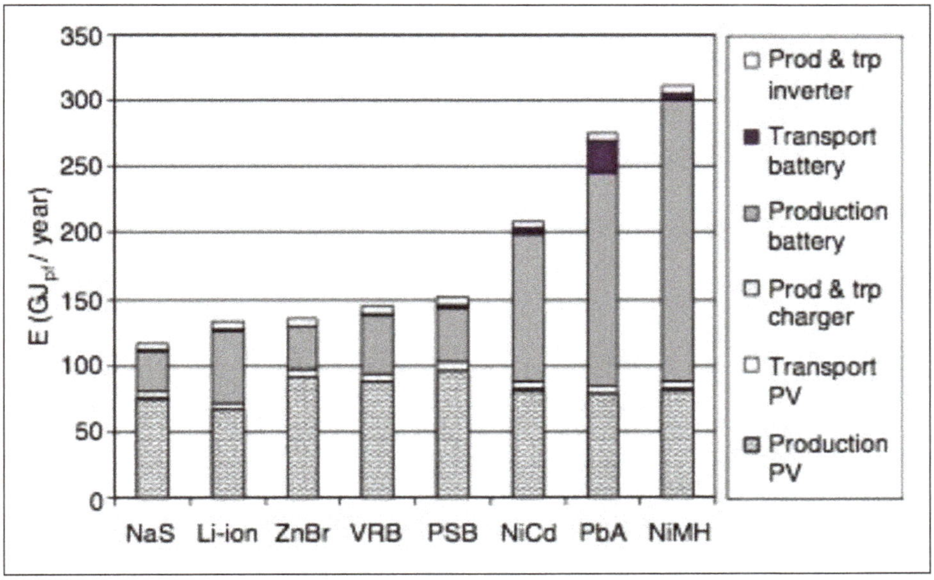

Energy requirements for production of the PV-battery system components expressed per kilogram component.

Influence of Different user Conditions

The consequences of recycling of battery materials is analysed in figure. When not using materials from recycled production, the energy requirement increases 17– 75 GJ/yr, which is 10–53% higher than if materials of recycled origin are used. The difference between virgin and recycled materials production is greatest for the ZnBr battery.

The energy intensity for plane transportation is 28 times higher than that for truck transportation. ZnBr is least affected by plane transport due to its long service life and high energy density.

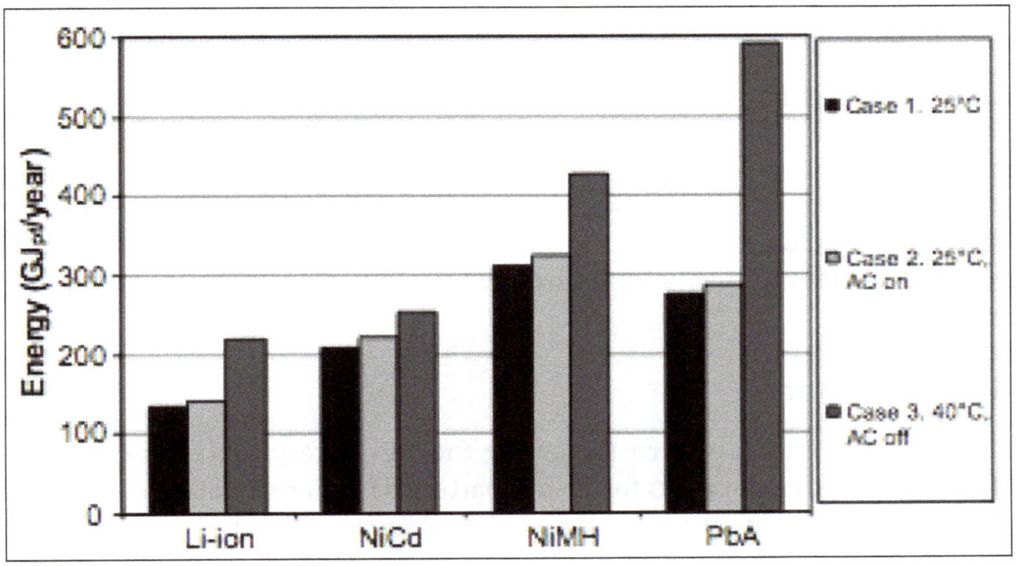

Energy requirements for production and transport of PV-battery systems at different operating temperatures.

The results show that for certain battery technologies and conditions, batteries contribute significantly to the total energy requirement for production of PV-battery systems.

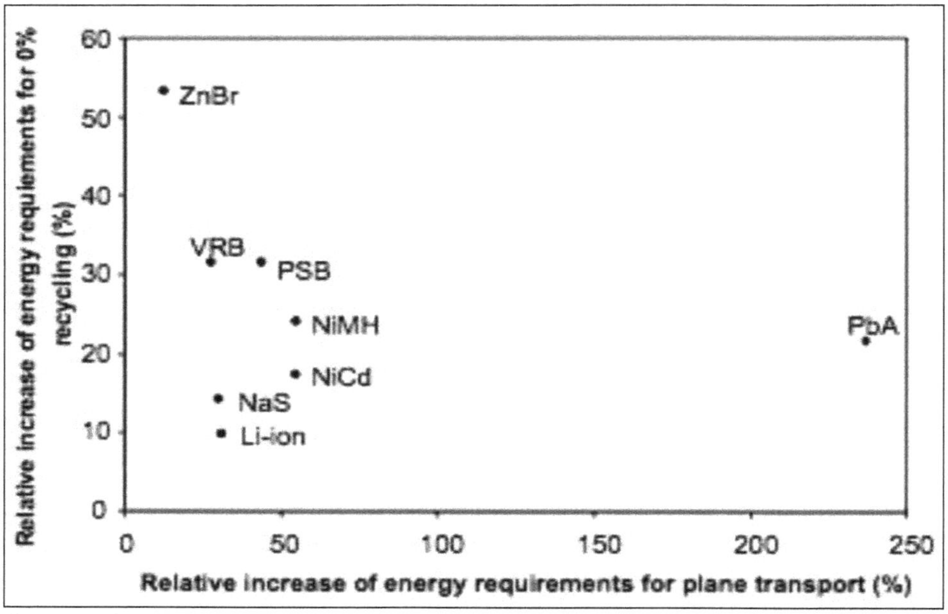

Changes of energy requirements for different recycling rates and modes of transportation.

Uncertainties

The low and high values indicate the uncertainties in the data as well as the improvement potential of different technologies. Input data with large impact and large uncertainty interval is the battery charge–discharge efficiency and the battery service life. The uncertainty in output results for different battery technologies vary between 8% and 61%. The difference between low and high values of input data is 1.1–2.2 times, where the highest variability is for NiMH and PbA. Since all battery technologies, except for PbA and NiCd, are immature for PV applications, there are uncertainties about their performance.

Only a small number of demonstration units have been built of VRB, PSB, ZnBr and NaS batteries, and mass production is likely to improve the production efficiency of these batteries. Energy requirements for production of batteries may vary considerably depending on material requirements and where and how they are manufactured. Uncertainties are due to restricted availability of information since companies manufacturing batteries protect their technology from competitors. Material requirements for immature technologies can change fast in the course of development.

Energy Return Factor

Figure shows that the energy return factor for the PV-battery system ranges from 2.2 for NiMH batteries (low estimate) to 10 for Li-ion batteries (high estimate). The NaS battery has the highest average energy return factor (8.5), which means that the PV–NaS battery system will replace 8.5 times more energy throughout its lifetime than the energy required for its production.

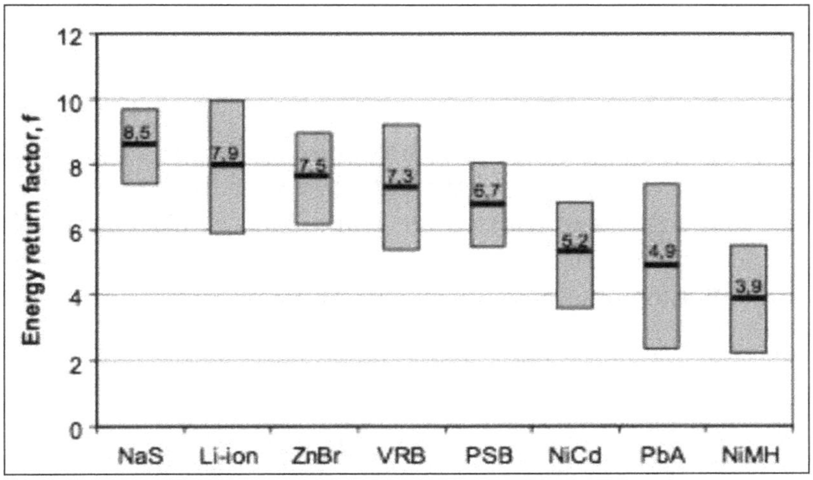

Energy return factors for the PV-battery systems.

The uncertainty intervals of many batteries are overlapping. This shows the importance of using case-specific data when comparing different battery technologies. The greatest uncertainty is found for systems with a PbA battery due to the high variability in lifetime and energy requirements. The low estimate of the energy return factor for PbA (using air transport) is as low as 0.68. If the operating temperature at the same time is increased to 40 °C and no material is recycled, then the energy return factor is 0.25. In this extreme case, four times more primary fossil energy is needed to produce and transport the system than what is replaced.

The production and transportation of batteries contributes 24–70% to the total indirect energy use of the PV-battery system compared to 26–68% for the PV array. Given a system lifetime of 30 years, the PV array contributes 1.8–3.3 yr and batteries 0.72–10 yr to the energy payback time depending on the technology, showing the energy related significance of batteries in PV- battery systems.

The importance of energy requirements for battery production increases in relation to PV at locations with high irradiation and vice versa. At locations with low irradiation levels, the power rating of the PV arrays has to be increased to produce the same amount of electricity. This results in higher indirect energy requirements for production of PV arrays. The importance of the level of solar irradiation for the total energy return factor depends on the relative contribution of the PV array to the indirect energy requirements of the PV-battery system. An increase of the solar irradiation by 30% increases the energy return factor of the PV- battery system by 9–20%.

The energy return factor is very sensitive to assumptions about the conversion efficiency of the electricity generation technology the PV system replaces. A conversion efficiency of 0.20 was selected as a default value to represent a Diesel generator since it is likely that a PV-battery system can replace its use in an off grid application. If the efficiency of the Diesel generator is 0.25, the energy return factor is reduced by 20% (1-0.20/0.25). Assuming that the PV-battery system replaces a grid connected system where the electricity conversion efficiency is 0.35, the energy return factor decreases by 43%.

The Overall Battery Efficiency

The direct efficiency of the battery system (charger, battery and inverter) is 0.50–0.85. Adding the indirect energy requirement results in overall efficiencies from 0.41 for the NiMH battery (low estimate) to 0.80 for the Li-ion battery (high estimate).

Figure below shows how the gross electricity requirement is used. The energy requirements for production and transport of the charger, the battery and the inverter have been converted to electricity. Energy losses in the batteries are significant and vary from 9% for the Li-ion battery to 27–33% for the redox flow batteries. Losses in the charger and inverter are 6–8% and 4–5% of the gross energy requirement, respectively. Production and transport of the charger and inverter contribute less than 1% of the gross energy requirement, while production and transport of batteries contribute 11–19% for NiCd, PbA and NiMH.

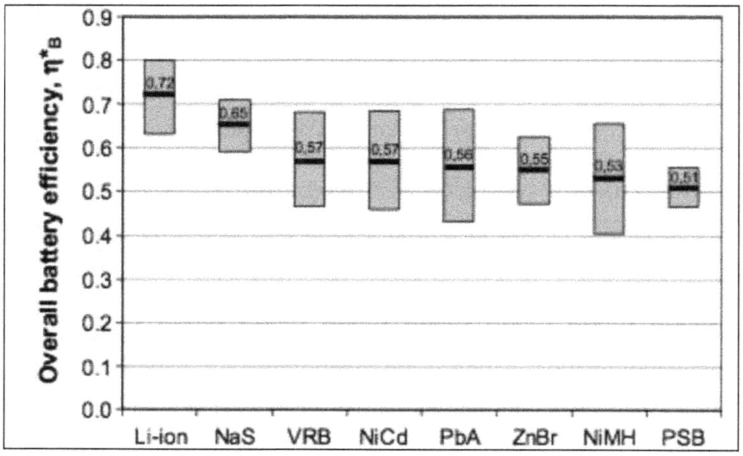

Overall battery efficiencies including production and transport of charger, battery and invertir.

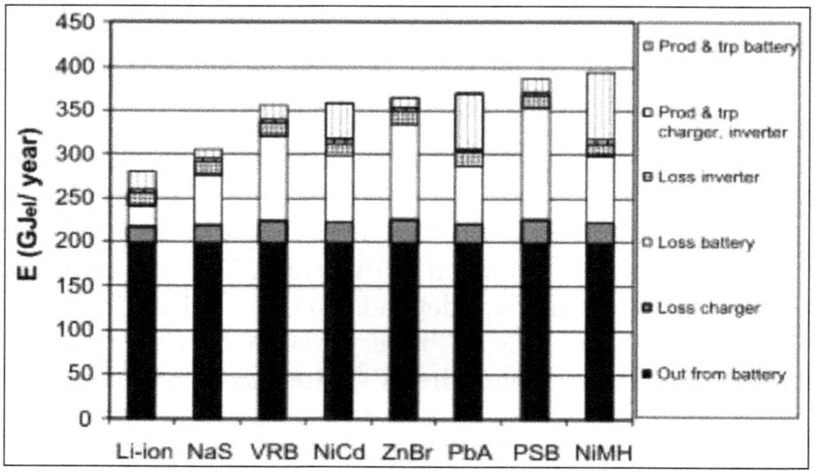

Gross electricity requirements for charger, battery and inverter including their production and transport.

Different Efficiency Measures and Battery Design

To increase energy efficiency, the technical performance of batteries may be improved. The relative importance of different battery parameters for energy efficiency was, therefore, analysed including service life, charge–discharge efficiency and energy requirements for production . For all batteries, the overall battery efficiency is most sensitive to changes of the charge-discharge efficiency. Changes in the efficiency of the Li-ion battery have a relatively small influence compared with the other battery technologies. This is because the losses in the Li-ion battery correspond to only 9% of the gross energy requirement, compared with 18–33% for the other Technologies.

Improvement in the charge–discharge efficiency is also most important for the energy return factor of Li-ion, NaS, PSB, VRB and ZnBr batteries. Since losses in the battery must be compensated by higher energy input, low battery efficiency results in a larger PV array and charger, which means higher indirect energy requirements.

The service life (and energy density) affects both energy requirements for production and transport. Short battery service life means that batteries have to be replaced more often, resulting in higher energy requirements for battery production and transport. The low energy density and short service life of PbA results in a larger share of transportation energy.

Uncertainties

The low and high values indicate the uncertainties in the results as well as the improvement potential of different technologies. Input data with high influence and large uncertainty interval is the battery charge–discharge efficiency and the battery service life. The uncertainty in output results for different battery technologies vary between 8% and 61%. The difference between low and high values of input data is 1.1–2.2 times, where the highest variability is for NiMH and PbA.

Since all battery technologies, except for PbA and NiCd, are immature for PV applications, the stated uncertainties for the other batteries could be underestimated.

The performance ratio, which is the ratio between the final yield and the reference yield of a PV system, is a common indicator for expressing the efficiency from PV array to useful energy.

It was calculated calculated the energy payback time for PbA batteries in a solar home system (SHS) to be 10–19 yr. The number of days of autonomy that was used for this calculation was 6.8–20 days (battery voltage 12 V, battery capacity 70–100 A h and array output 60–124 W h/day). Recalculated to 3 days of autonomy, the energy payback time for the battery is 1.5–8.4 yr. The service life for the starting- lighting-ignition (SLI) battery is assumed to be 3 yr. In SHS, without charge regulator and poor battery maintenance, the service life may be less than 3 yr when standard SLI- PbA batteries are used.

Improvement of the charge/discharge strategy can extend the service life of PbA batteries in PV applications.

Solar Inverter

A solar inverter or PV inverter, is a type of electrical converter which converts the variable direct current (DC) output of a photovoltaic (PV) solar panel into a utility frequency alternating current (AC) that can be fed into a commercial electrical grid or used by a local, off-grid electrical network. It is a critical balance of system (BOS)–component in a photovoltaic system, allowing the use of ordinary AC-powered equipment. Solar power inverters have special functions adapted for use with photovoltaic arrays, including maximum power point tracking and anti-islanding protection.

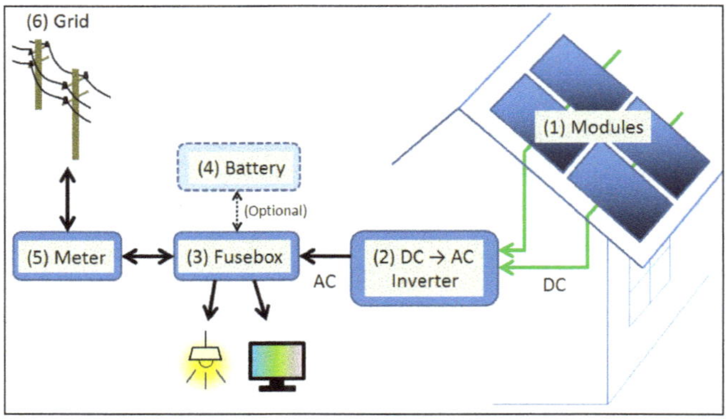

Simplified schematics of a grid-connected residential photovoltaic power system.

Classification

Solar inverters may be classified into three broad types:

- Stand-alone inverters, used in isolated systems where the inverter draws its DC energy from batteries charged by photovoltaic arrays. Many stand-alone inverters also incorporate integral battery chargers to replenish the battery from an AC source, when available. Normally these do not interface in any way with the utility grid, and as such, are not required to have anti-islanding protection.
- Grid-tie inverters, which match phase with a utility-supplied sine wave. Grid-tie inverters are designed to shut down automatically upon loss of utility supply, for safety reasons. They do not provide backup power during utility outages.
- Battery backup inverters, are special inverters which are designed to draw energy from a battery, manage the battery charge via an onboard charger, and export excess energy to the utility grid. These inverters are capable of supplying AC energy to selected loads during a utility outage, and are required to have anti-islanding protection.
- Intelligent hybrid inverters, manage photovoltaic array, battery storage and utility grid, which are all coupled directly to the unit. These modern all-in-one systems are usually highly versatile and can be used for grid-tie, stand-alone or backup applications but their primary function is self-consumption with the use of storage.

Maximum Power Point Tracking

Solar inverters use maximum power point tracking (MPPT) to get the maximum possible power from the PV array. Solar cells have a complex relationship between solar irradiation, temperature and total resistance that produces a non-linear output efficiency known as the I-V curve. It is the purpose of the MPPT system to sample the output of the cells and determine a resistance (load) to obtain maximum power for any given environmental conditions.

The fill factor, more commonly known by its abbreviation FF, is a parameter which, in conjunction with the open circuit voltage (V_{oc}) and short circuit current (I_{sc}) of the panel, determines the maximum power from a solar cell. Fill factor is defined as the ratio of the maximum power from the solar cell to the product of V_{oc} and I_{sc}.

There are three main types of MPPT algorithms: perturb-and-observe, incremental conductance and constant voltage. The first two methods are often referred to as hill climbing methods; they rely on the curve of power plotted against voltage rising to the left of the maximum power point, and falling on the right.

Solar Micro-inverters

A solar micro-inverter in the process of being installed. The ground wire is attached to the lug and the panel's DC connections are attached to the cables on the lower right. The AC parallel trunk cable runs at the top.

Solar micro-inverter is an inverter designed to operate with a single PV module. The micro-inverter converts the direct current output from each panel into alternating current. Its design allows parallel connection of multiple, independent units in a modular way.

Micro-inverter advantages include single panel power optimization, independent operation of each panel, plug-and play installation, improved installation and fire safety, minimized costs with system design and stock minimization.

A 2011 study at Appalachian State University reports that individual integrated inverter setup yielded about 20% more power in unshaded conditions and 27% more power in shaded conditions compared to string connected setup using one inverter. Both setups used identical solar panels.

Grid Tied Solar Inverters

Solar grid-tie inverters are designed to quickly disconnect from the grid if the utility grid goes down. This is an NEC requirement that ensures that in the event of a blackout, the grid tie inverter will shut down to prevent the energy it produces from harming any line workers who are sent to fix the power grid.

Grid-tie inverters that are available on the market today use a number of different technologies. The inverters may use the newer high-frequency transformers, conventional low-frequency transformers, or no transformer. Instead of converting direct current directly to 120 or 240 volts AC, high-frequency transformers employ a computerized multi-step process that involves converting the power to high-frequency AC and then back to DC and then to the final AC output voltage.

Historically, there have been concerns about having transformerless electrical systems feed into the public utility grid. The concerns stem from the fact that there is a lack of galvanic isolation between the DC and AC circuits, which could allow the passage of dangerous DC faults to the AC side. Since 2005, the NFPA's NEC allows transformer-less (or non-galvanically) inverters. The VDE 0126-1-1 and IEC 6210 also have been amended to allow and define the safety mechanisms needed for such systems. Primarily, residual or ground current detection is used to detect possible fault conditions. Also isolation tests are performed to ensure DC to AC separation.

Many solar inverters are designed to be connected to a utility grid, and will not operate when they do not detect the presence of the grid. They contain special circuitry to precisely match the voltage, frequency and phase of the grid.

Solar Pumping Inverters

Advanced solar pumping inverters convert DC voltage from the solar array into AC voltage to drive submersible pumps directly without the need for batteries or other energy storage devices. By utilizing MPPT (maximum power point tracking), solar pumping inverters regulate output frequency to control the speed of the pumps in order to save the pump motor from damage.

Solar pumping inverters usually have multiple ports to allow the input of DC current generated by PV arrays, one port to allow the output of AC voltage, and a further port for input from a water-level sensor.

Market

As of 2014, conversion efficiency for state-of-the-art solar converters reached more than 98 percent. While string inverters are used in residential to medium-sized commercial PV systems, central inverters cover the large commercial and utility-scale market. Market-share for central and string inverters are about 50 percent and 48 percent, respectively, leaving less than 2 percent to micro-inverters.

TYPES OF SOLAR ELECTRIC SYSTEMS

Stand-alone Power System

A stand-alone power system (SAPS or SPS), also known as remote area power supply (RAPS), is an off-the-grid electricity system for locations that are not fitted with an electricity distribution system. Typical SAPS include one or more methods of electricity generation, energy storage, and regulation.

Electricity is typically generated by one or more of the following methods:

- Photovoltaic system using solar panels.
- Wind turbine.

- Geothermal source.
- Micro combined heat and power.
- Micro hydro.
- Diesel or biofuel generator.
- Thermoelectric generator (TEGs).

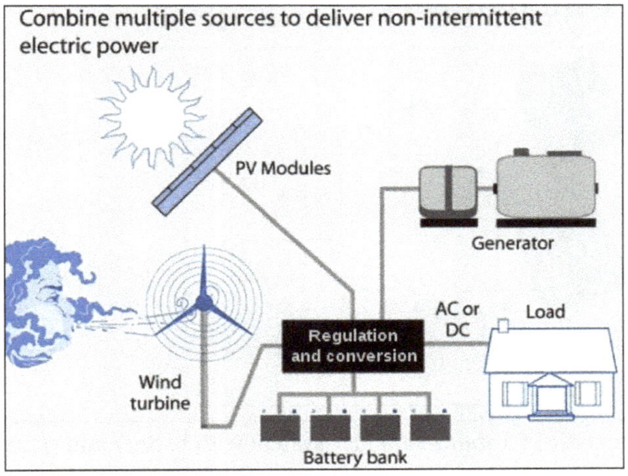

Schematics of a hybrid system.

Storage is typically implemented as a battery bank, but other solutions exist including fuel cells. Power drawn directly from the battery will be direct current extra low voltage (DC ELV), and this is used especially for lighting as well as for DC appliances. An inverter is used to generate AC low voltage, which more typical appliances can be used with.

A typical stand-alone solar PV system at a sewage treatment plant in Santuari de Lluc, Spain.

Stand-alone photovoltaic power systems are independent of the utility grid and may use solar panels only or may be used in conjunction with a diesel generator, a wind turbine or batteries.

Types

The two types of stand-alone photovoltaic power systems are direct-coupled system without batteries and stand alone system with batteries.

Direct-coupled System

The basic model of a direct coupled system consists of a solar panel connected directly to a dc load. As there are no battery banks in this setup, energy is not stored and hence it is capable of powering common appliances like fans, pumps etc. only during the day. MPPTs are generally used to efficiently utilize the Sun's energy especially for electrical loads like positive-displacement water pumps. Impedance matching is also considered as a design criterion in direct-coupled systems.

Stand Alone System with Batteries

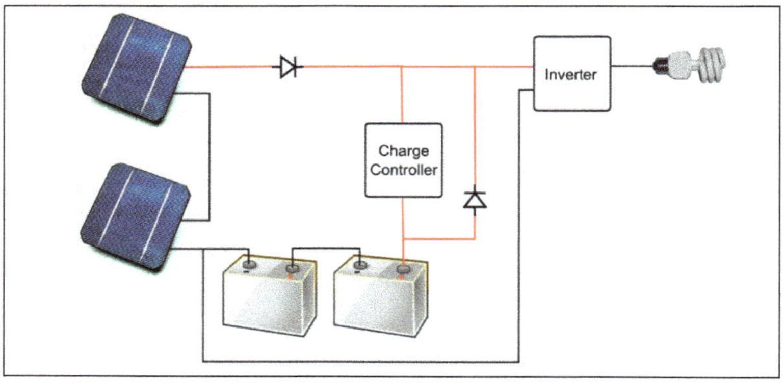

Schematic of a stand-alone PV system with battery and charger.

In stand-alone photovoltaic power systems, the electrical energy produced by the photovoltaic panels cannot always be used directly. As the demand from the load does not always equal the solar panel capacity, battery banks are generally used. The primary functions of a storage battery in a stand-alone PV system are:

- Energy Storage Capacity and Autonomy: To store energy when there is an excess is available and to provide it when required.
- Voltage and Current Stabilization: To provide stable current and voltage by eradicating transients.
- Supply Surge Currents: to provide surge currents to loads like motors when required.

Hybrid System

The hybrid power plant is a complete electrical power supply system that can be easily configured to meet a broad range of remote power needs. There are three basic elements to the system - the power source, the battery, and the power management center. Sources for hybrid power include wind turbines, diesel engine generators, thermoelectric generators and solar PV systems. The battery allows autonomous operation by compensating for the difference between power production and use. The power management center regulates power production from each of the sources, controls power use by classifying loads, and protects the battery from service extremes.

System Monitoring

Monitoring photovoltaic systems can provide useful information about their operation and what should be done to improve performance, but if the data are not reported properly, the effort is wasted. To be helpful, a monitoring report must provide information on the relevant aspects of the operation in terms that are easily understood by a third party. Appropriate performance parameters need to be selected, and their values consistently updated with each new issue of the report. In some cases it may be beneficial to monitor the performance of individual components in order to refine and improve system performance, or be alerted to loss of performance in time for preventative action. For example, monitoring battery charge/discharge profiles will signal when replacement is due before downtime from system failure is experienced.

IEC Standard 61724

IEC has provided a set of monitoring standards called the "Standard for Photovoltaic system performance monitoring". It focuses on the photovoltaic system's electrical performance and it does not address hybrids or prescribe a method for ensuring that performance assessments are equitable.

Performance Assessment

Performance assessment involves:

- Data collection, which is a straightforward process of measuring parameters.
- Evaluation of that data in a manner that provides useful information.
- Dissemination of useful information to the end user.

Load Related Problems

The wide range of load related problems identified are classified into the following types:

- Wrong selection: Some loads cannot be used with stand-alone PV systems.
- House wiring: Inadequate or low quality wiring and protection devices can affect the system's response.
- Low efficiency: Low efficiency loads may increase energy consumption.
- Stand-by loads: Stand-by mode of some loads waste energy.
- Start-up: High current drawn by some loads during start-up Current spikes during the start-up can overload the system temporarily.
- Reactive power: The circulating current can differ from the current consumed when capacitive or inductive loads are used.
- Harmonic distortion: Non-linear loads may create distortion of the inverter waveform.
- Mismatch between load and inverter size: When a higher rated inverter is used for a lower capacity load, overall efficiency is reduced.

Grid-connected Photovoltaic Power System

A grid-connected photovoltaic system, or grid-connected PV system is an electricity generating solar PV power system that is connected to the utility grid. A grid-connected PV system consists of solar panels, one or several inverters, a power conditioning unit and grid connection equipment. They range from small residential and commercial rooftop systems to large utility-scale solar power stations. Unlike stand-alone power systems, a grid-connected system rarely includes an integrated battery solution, as they are still very expensive. When conditions are right, the grid-connected PV system supplies the excess power, beyond consumption by the connected load, to the utility grid.

Operation

Photovoltaic power station at Nellis Air Force Base, United States.

Residential, grid-connected rooftop systems which have a capacity more than 10 kilowatts can meet the load of most consumers. They can feed excess power to the grid where it is consumed by other users. The feedback is done through a meter to monitor power transferred. Photovoltaic wattage may be less than average consumption, in which case the consumer will continue to purchase grid energy, but a lesser amount than previously. If photovoltaic wattage substantially exceeds average consumption, the energy produced by the panels will be much in excess of the demand. In this case, the excess power can yield revenue by selling it to the grid. Depending on their agreement with their local grid energy company, the consumer only needs to pay the cost of electricity consumed less the value of electricity generated. This will be a negative number if more electricity is generated than consumed. Additionally, in some cases, cash incentives are paid from the grid operator to the consumer.

Connection of the photovoltaic power system can be done only through an interconnection agreement between the consumer and the utility company. The agreement details the various safety standards to be followed during the connection.

Features

Electric power from photovoltaic panels must be converted to alternating current by a power inverter if it is intended for delivery to a power grid. The inverter sits between the solar array and the grid, and may be a large stand-alone unit or may be a collection of small inverters attached to individual solar panels as an AC module. The inverter must monitor grid voltage, waveform, and frequency. The inverter must detect failure of the grid supply and must not supply power to the grid.

An inverter connected to a malfunctioning power line will automatically disconnect in accordance with safety rules, which vary by jurisdiction. The location of the fault current plays a crucial part in deciding whether the protection mechanism of the inverter will kick in, especially for low and medium electricity supply network. A protection system must ensure proper operation for faults external to the inverter on the supply network. The inverter must be designed to synchoronize its AC frequency with the grid, to ensure correct direction of power flow.

Anti-islanding

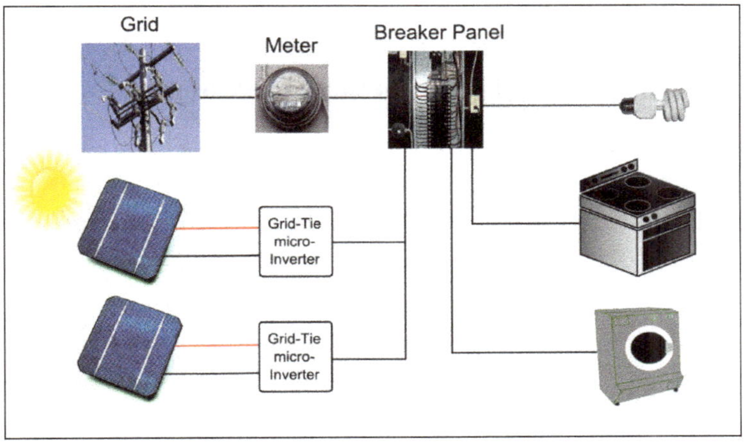

Diagram of a residential grid-connected PV system.

Islanding is the condition in which a distributed generator continues to power a location even though power from the electric utility grid is no longer present. Islanding can be dangerous to utility workers, who may not realize that a circuit is still powered, even though there is no power from the electrical grid. For that reason, distributed generators must detect islanding and immediately stop producing power; this is referred to as anti-islanding.

In the case of a utility blackout in a grid-connected PV system, the solar panels will continue to deliver power as long as the sun is shining. In this case, the supply line becomes an "island" with power surrounded by a "sea" of unpowered lines. For this reason, solar inverters that are designed to supply power to the grid are generally required to have automatic anti-islanding circuitry in them. In intentional islanding, the generator disconnects from the grid, and forces the distributed generator to power the local circuit. This is often used as a power backup system for buildings that normally sell their power to the grid.

There are two types of anti-islanding control techniques:

- Passive: The voltage and/or the frequency change during the grid failure is measured and a positive feedback loop is employed to push the voltage and/or the frequency further away from its nominal value. Frequency or voltage may not change if the load matches very well with the inverter output or the load has a very high quality factor (reactive to real power ratio). So there exists some Non Detection Zone (NDZ).
- Active: This method employs injecting some error in frequency or voltage. When grid fails, the error accumulates and pushes the voltage and/or frequency beyond the acceptable range.

Advantages

- Systems such as Net Metering and Feed-in Tariff which are offered by some system operators, can offset a customers electricity usage costs. In some locations though, grid technologies cannot cope with distributed generation feeding into the grid, so the export of surplus electricity is not possible and that surplus is earthed.
- Grid-connected PV systems are comparatively easier to install as they do not require a battery system.
- Grid interconnection of photovoltaic (PV) power generation systems has the advantage of effective utilization of generated power because there are no storage losses involved.
- A photovoltaic power system is carbon negative over its lifespan, as any energy produced over and above that to build the panel initially offsets the need for burning fossil fuels. Even though the sun doesn't always shine, any installation gives a reasonably predictable average reduction in carbon consumption.

Disadvantages

- Grid-connected PV can cause issues with voltage regulation. The traditional grid operates under the assumption of one-way, or radial, flow. But electricity injected into the grid increases voltage, and can drive levels outside the acceptable bandwidth of ±5%.
- Grid-connected PV can compromise power quality. PV's intermittent nature means rapid changes in voltage. This not only wears out voltage regulators due to frequent adjusting, but also can result in voltage flicker.
- Connecting to the grid poses many protection-related challenges. In addition to islanding, as mentioned above, too high levels of grid-connected PV result in problems like relay desensitization, nuisance tripping, interference with automatic reclosers, and ferroresonance.

References

- Hohm, D. P.; Ropp, M. E. (2003). "Comparative Study of Maximum Power Point Tracking Algorithms". Progress in Photovoltaics: Research and Applications. 11: 47–62. Doi:10.1002/pip.459
- Solar-energy-power-generation: electronicshub.org, Retrieved 28 January, 2019
- "Invert your thinking: Squeezing more power out of your solar panels". Scientificamerican.com. Retrieved 2011-06-09
- What-are-solar-panels: solar-facts-and-advice.com, Retrieved 14 February, 2019
- Badwal, Sukhvinder P. S.; Giddey, Sarbjit S.; Munnings, Christopher; Bhatt, Anand I.; Hollenkamp, Anthony F. (24 September 2014). "Emerging electrochemical energy conversion and storage technologies". Frontiers in Chemistry. 2. Doi:10.3389/fchem.2014.00079. PMC 4174133. PMID 25309898
- Design-of-Solar-Power-Inverter: researchgate.net, Retrieved 29 March, 2019
- "Guidelines for Monitoring Stand-Alone Photovoltaic Systems: Methodology and Equipment". Iea-pvps.org. Retrieved 2011-07-21

Applications of Solar Power

CHAPTER 5

Solar power is used for various purposes such as in solar water heating, solar pool heating, solar cookers, solar street lighting, solar-powered stirling engine, solar furnaces, solar inverters, solar chimneys, etc. The chapter closely examines these key applications of solar power to provide an extensive understanding of the subject.

SOLAR PANELS

Solar panels are active solar devices that convert sunlight into electricity. They come in a variety of rectangular shapes and are usually installed in combination to produce electricity.

Here is an image which will help you better understand some basic terminology.

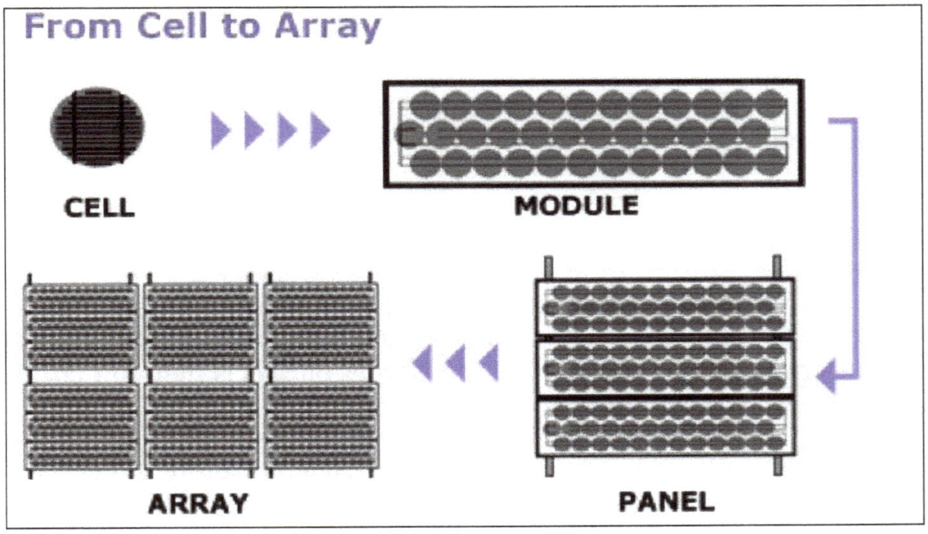

- Solar Cell: Semiconductor device that converts sunlight into direct current (DC) electricity.
- Module: PV modules consist of PV cell circuits sealed in an environmentally protective laminate and are the fundamental building block of PV systems.
- Solar Panel: Includes one or more PV modules assembled as a pre-wired, field-installable unit.
- Array: A PV array is the complete power-generating unit, consisting of any number of PV modules and panels.

Where are Solar Panel's Used?

The following is a breakdown (based on installations in the USA) showing where solar panels are used.

What are Solar Panels Made of?

The primary component of a solar panel is the solar cells, or photovoltaic cell. This is the key component that converts sunlight into electricity.

At the present time about 80% of all solar panels are made from crystalline silicon (i.e., monocrystalline, polycrystalline, amorphous silicon, or hybrids) solar cells. Typically the solar cells are laid out in a grid pattern – with perhaps as many as 72 different solar cells.

The other 20% consist primarily of solar cells made mostly from Cadmium Telluride and a small but growing amount from CIGS. The appeal of these types of cells is their low cost resulting from the fact they can be made in large single sheets.

The solar panels, after usually after being hermetically sealed to protect them are covered in a non-reflective glass to protect the solar cells from environmental damage and placed into a rigid frame.

Typically, the frame is designed to prevent it from deforming due to freezing weather or strong winds. The frame will usually include a drainage hole to help prevent water buildup on the panels, which can reduce output.

The back of the panel is also sealed to prevent damage. This is typically where the junction box is located.

Tip: Several companies have begun embedding electronics into PV junction boxes. This enables performing Maximum Power Point Tracking (MPPT) for each module individually, and the measurement of performance data for monitoring and fault detection at module level. Some of these solutions make use of Power Optimizers, a DC to DC converter technology developed to maximize the power harvest from solar photovoltaic systems.

How much Power do Solar Panels Produce?

Each manufacturer is required to provide a datasheet, which you can typically download from the company's website.

Using standardized testing protocols (to make comparison between solar panels possible) each manufacturer will list the peak output of the panel (or kWp as it is abbreviated in Europe).

Unless the power is needed to charge a battery – an inverter converts the DC electricity to AC electricity that can be used in the home or sold back to the grid.

External Factors that will Effect Amount of Electricity Produced

The amount of power solar panels produce is influenced by the amount of sunlight falling on your specific location per year, the efficiency of the underlying solar cell technology, the materials and technology used in making the solar panel, keeping your panels clean, and the amount of time the solar panel has been in use.

When purchasing solar panels, it is therefore wise to look beyond size and look at the dollars/watt ratio. As solar project developers in Europe - we look more at the estimated kWh produced over the life of the solar panel and calculate what our net return on investment will be.

To learn more about the basic factors that affect the amount of power a solar panel will produce – read some of our other posts, namely:

- Irradiance – How much solar energy reaches the earth.
- Insolation – How much solar energy reaches each spot on the earth.
- Heat – And how this can lower the amount of power produced.
- Cleanliness – And how this can lower the amount of power produced.
- kWh – How to calculate expected solar power production/per year.
- Duration - Solar panels are usually rated for 25 years – but can last much longer, however solar cells become less efficient over time (each type of solar cell technology has a different life span – with monocrystalline cells lasting the longest and the thin-film technologies – being the shortest).

Tip: In many cases the solar panels will last longer than the roof – so this is something to consider on a large installation.

Tip: Before buying a solar panel – read the performance warranty. These vary by company – and provide you with a good way to see how much confidence the manufacturer has in their own quality control (i.e., better made panels have better performance warranties).

When running your calculations to see how much electricity your panels will produce use the amount that is guaranteed (not the listed peak power) – it's a bit more conservative – but a more reliable estimate in the long run.

SOLAR-POWERED STIRLING ENGINE

10 kW Dish-stirling System.

A solar powered Stirling engine is a heat engine powered by a temperature gradient generated by the sun. It was patented by Roelf J. Meijer in 1987. His invention relates a heat engine, such as a Stirling cycle engine, with a solar dish collector in order to produce electricity. This apparatus consists of a large dish aimed at the sun to reflect the rays into the focus point, which is located at the center of the dish. Solar energy is then collected in the form of heat to fuel a Stirling cycle engine, which operates by letting heat flow from a hot source to a cold sink in order to do work. The work output of the Stirling cycle is then used to drive a generator and create electric power. Moreover, in Meijer's solar-powered engine, it is important that the dish always be pointed directly at the sun so that no shadows would be present in the solar dish collector, hence optimizing heat collection. This is where he ran into some issues because, in order for the apparatus to have a complete range of motion, lubrication and rotational systems would be necessary and may compromise structural stability.

A second type of solar-powered Stirling engine was patented by NASA on August 3, 1976 which employed the use of solar energy in order to freely pump water from a river, lake, or stream. The purpose of this apparatus is to "provide a low-cost, low-technology pump having particular utility in irrigation systems employed in underdeveloped arid regions of the earth using the basic principles of the Stirling heat engine.

Recently, a company called Sunvention Solar Energy created a device very similar to what NASA came up with in 1976 that can pump 100,000 gallons per day purely off of solar energy and usage of the Stirling cycle costing only US$1,250. This apparatus, much like the others, uses a large solar dish to collect heat from the sun creating a high temperature source and also employs the low temperature water it collects from a nearby stream as its low temperature source allowing for a great range in temperature which in turn allows for more work to be done The work done in Sunvention's apparatus is used to pump the water into nearby crop fields allowing for a "low-cost, low-technology pump having particular utility in irrigation systems employed in underdeveloped arid regions of the earth".

SOLAR STREET LIGHT

Solar street lights are raised light sources which are powered by solar panels generally mounted on the lighting structure or integrated in the pole itself. The solar panels charge a rechargeable battery, which powers a fluorescent or LED lamp during the night.

Features

Most solar lights turn on and turn off automatically by sensing outdoor light using solar panel voltage. Solar streetlights are designed to work throughout the night. Many can stay lit for more than one night if the sun is not in the sky for an extended period of time. Older models included lamps that were not fluorescent or LED. Solar lights installed in windy regions are generally equipped with flat panels to better cope with the winds.

Modern designs use wireless technology and fuzzy control theory for battery management. The street lights using this technology can operate as a network with each light having the capability of performing the turning on and off of the network.

Components

Solar street lights consist of four main parts:

Solar Panel

The solar panel is one of the most important parts of a solar street light, as the solar panel is able to convert solar energy into electricity that the lamps can use. There are two types of solar panel commonly used in solar street lights: mono-crystalline and poly-crystalline. The conversion rate of mono-crystalline solar panels are much higher than their poly-crystalline counterpart. Solar panels also vary in wattage systems.

Lighting Fixture

LEDs are usually used as the primary lighting source of modern solar street lights, as the LED will provide much higher luminosity with lower energy consumption. The energy consumption of an LED fixture is at least 50% lower than the HPS fixture counterpart which is widely used as the lighting source in traditional street lights. A lack of warm-up time in LEDs also allows for use of motion detectors for additional efficiency gains.

Rechargeable Battery

Batteries will store the electricity generated by the solar panel during the day and provide energy to the fixture during night. The life cycle of the battery is very important to the lifetime of the light and the capacity of the battery will affect the backup days of the lights. There are two types of batteries commonly used in solar-powered street lights- gel cell deep cycle batteries as well as lead acid batteries. Lithium-ion batteries are also popular these days as they are compact in size and not prone to theft due to their lack of use in other situations unlike lead acid batteries.

Pole

Strong poles are necessary to all street lights, especially to solar street lights as there are often components mounted on the top of the pole: fixtures, panels and sometimes batteries. However, in some newer designs, the PV panels and all electronics are integrated in the pole itself. Wind resistance is also a factor.

In addition, accessories do exist for these types of poles, such as a foundation cage and a battery box.

Types

Solar street light at the bus stop.

Each street light can have its own photo voltaic panel, independent of other street lights. Alternately, a number of panels can be installed as a central power source on a separate location and supply power to a number of street lights.

All-in-one type solar street lights are gaining popularity due to their compact design which incorporates all of the parts necessary in a compact manner including the battery.

The city of Las Vegas, Nevada was the first city in the world that tested new EnGoPlanet Solar Street lights which are coupled with kinetic tiles that produce electricity when people walk over them.

Advantages

- Solar street lights are independent of the utility grid. Hence, the operation costs are minimized.
- Solar street lights require much less maintenance compared to conventional street lights.
- Since external wires are eliminated, risk of accidents are minimized.
- Electricity produced from solar panels is non-pollutive.
- Separate parts of a solar panel system can easily be transported.
- Energy costs can be saved.

Disadvantages

- Initial investment is higher compared to conventional street lights.
- Risk of theft is higher as equipment costs are comparatively higher.
- Snow or dust, combined with moisture can accumulate on horizontal PV-panels and reduce or even stop energy production.
- Rechargeable batteries will need to be replaced several times over the lifetime of the fixtures adding to the total lifetime cost of the light.

The charge and discharge cycles of the battery are also very important considering the overall cost of the project.

SOLAR LAMP

A solar lamp also known as solar light or solar lantern, is a lighting system composed of an LED lamp, solar panels, battery, charge controller and there may also be an inverter. The lamp operates on electricity from batteries, charged through the use of solar photovoltaic panel.

A solar lamp in Rizal Park, Philippines.

A garden solar lamp.

Solar-powered household lighting can replace other light sources like candles or kerosene lamps. Solar lamps have a lower operating cost than kerosene lamps because renewable energy from the sun is free, unlike fuel. In addition, solar lamps produce no indoor air pollution unlike kerosene lamps. However, solar lamps generally have a higher initial cost, and are weather dependent.

Solar lamps for use in rural situations often have the capability of providing a supply of electricity for other devices, such as for charging cell phones. American investors have been working towards developing a $10/unit solar lantern for replacement of kerosene lamps.

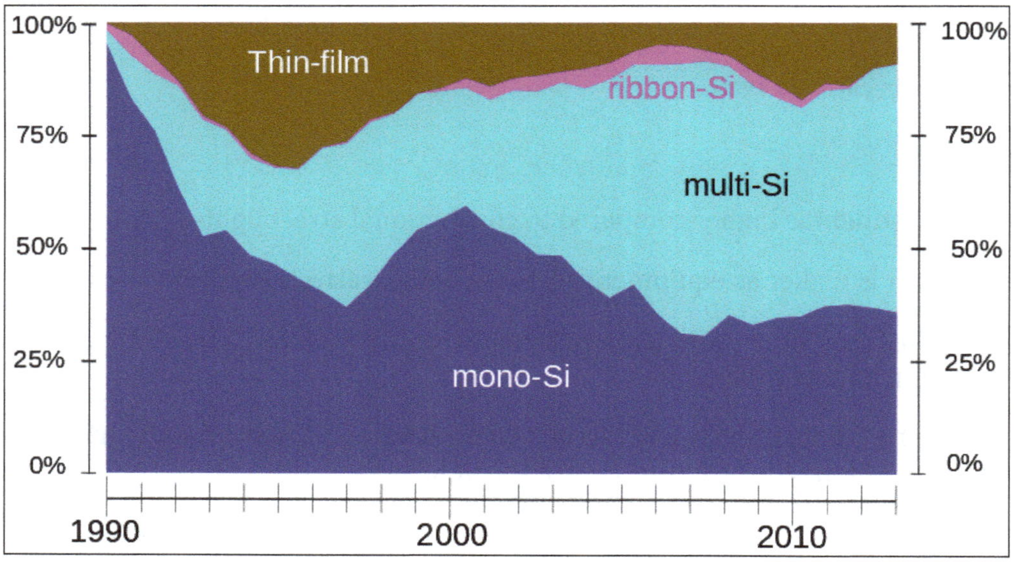

Global market-share in terms of annual production by PV technology since 1990.

Components

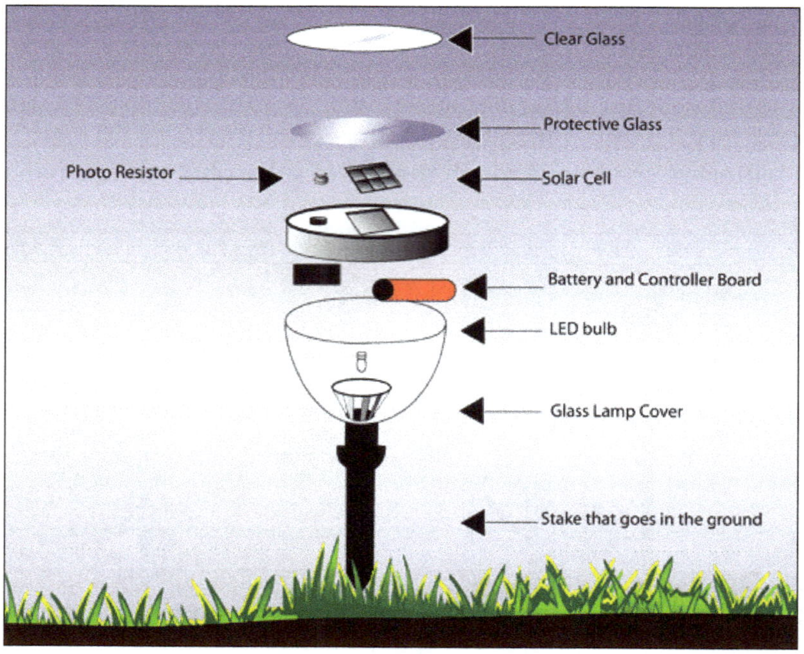

The whole structure of solar lamp.

Solar Panels

Most solar panels are made out of single crystalline silicon, a semiconductor material. A solar cell has two different layers of silicon. The lower layer has less electrons and hence has a slight positive charge due to the negative charge nature of electrons. In addition, the upper layer has more electrons and has slightly negative charge. A potential barrier is created between these two layers.

When the stream of light particles called photons enter, they give up their energy to the atoms in the silicon. It promotes one electron from a covalent bond to a next energy level from upper layer to the lower layer. This promotion of an electron allows freer movement within the crystal which produces a current. More light shines through, more electrons move around hence more current flows between. This process is called photovoltaic and photoelectric effect. Photovoltaic systems directly convert sunlight into electricity.

Solar panels are made out of layers of different materials, as you can see in below, in order of glass, encapsulate, crystalline cells, encapsulate, back sheet, junction box and lastly frame. The encapsulate keeps out moisture and contaminants which could cause problems.

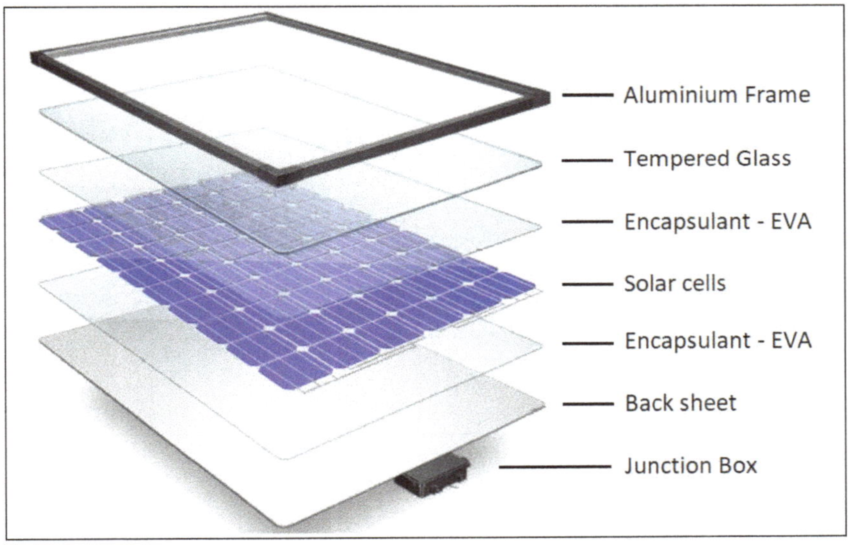

Battery

A battery is usually housed within a metal or plastic case. Inside the case are electrodes including cathodes and anodes where chemical reactions occur. A separator also exists between cathode and anode which stops the electrodes reacting together at the same time as allowing electrical charge to flow freely between the two. Lastly, the collector conducts a charge from the battery to outside.

Batteries inside solar lamps usually use gel electrolyte technology with high performance in deep discharging, in order to enable use in extreme ranges of temperature. It may also use lead-acid, nickel metal hydride, nickel cadmium, or lithium.

This part of the lamp saves up energy from the solar panel and provides power when needed at night when there is no light energy available.

In general, the efficiency of photovoltaic energy conversion is limited for physical reasons. Around 24% of solar radiation of a long wavelength is not absorbed. 33% is heat lost to surroundings, and further losses are of approximately 15-20%. Only 23% is absorbed which means a battery is a crucial part of solar lamp.

Charge Controller

Charge Controller controls the entire working systems to protect battery charge. It ensures, under any circumstances including extreme weather conditions with large temperature difference, the battery does not overcharge or over discharge and damage the battery even further.

This also includes additional parts such as light controller, time controller, sound, temperature compensation, lighting protection, reverse polarity protection and AC transfer switches which ensure sensitive back-up loads work normally when outage occurs.

Working Principles

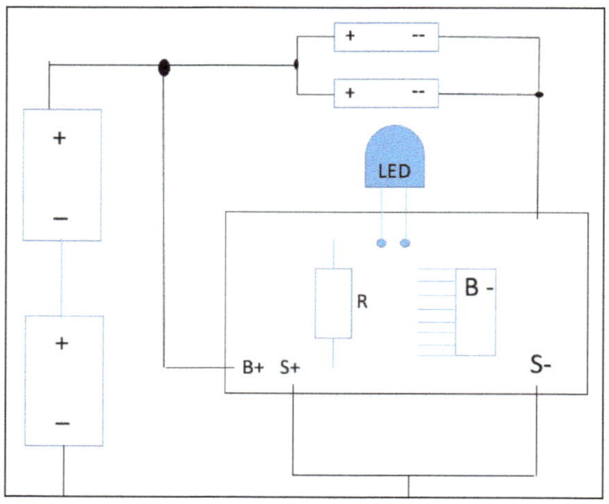

LED lights are used due to their high luminous efficiency and long life. Under the control of a DC charge controller, non-contact control automatically turns on the light at dark and switches off at daytime. It sometimes also combines with time controllers to set curtain time for it to automatically switch light on and off.

As shown in figure, the chip includes microchip(R), B-, B+, S- and S+. S+ and S- are both connected to solar panels with wire, one of which has plus charge and the other minus charge. B- and B+ are attached to two batteries in this case. The light will be shown through the LED light when all of these are connected.

Benefits

Solar lamps can be easier for customers to install and maintain as they do not require an electricity cable. Solar lamps can benefit owners with reduced maintenance cost and costs of electricity bills. Solar lamps can also be used in areas where there is no electrical grid or remote areas that lack a reliable electricity supply. There are many stories of people with lung disease, eye deterioration, burns and sometimes even death simply because they do not have a healthy alternative to light at night. Women have felt unsafe walking to the toilet outside after dark. Babies are being delivered by midwives using only a candle, and students cannot study when the sun goes down for lack of light leading to increased illiteracy and perpetual poverty. These are the realities for over 1 billion people around the globe. Lack of lighting equates to continued poverty felt around the world. In solar energy lamp it converts light energy into electrical energy, i.e. It is convenient to our daily life.

Solar energy output is limited by weather and can be less effective if it is cloudy, wet, or winter.

Households switching to solar lamps from kerosene lamps also gain from health risk associated with kerosene emissions. Kerosene often has negative impacts on human lungs.

The use of solar energy minimizes the creation pollution indoors, where kerosene have been linked to cases of health issues. However, photovoltaic panels are made out of silicon and other toxic metals including lead that can be difficult to dispose of.

The use of solar lights improves education for students who live in households without electricity. When the nonprofit Unite-To-Light donated solar-lamps to schools a remote region of Kwa Zulu Natal in South Africa test scores and pass rates improved by over 30%. The light gives students added time to study after dark.

A 2017 experimental study in unelectrified areas of northern Bangladesh found that the use of solar lanterns decreased total household expenditure, increased children's home-study hours and increased school attendance. It did not however improve the children's educational achievement to any large extent.

Use

Solar Street Light

These lights provide a convenient and cost-effective way to light streets at night without the need of AC electrical grids for pedestrians and drivers. They may have individual panels for each lamp of a system, or may have a large central solar panel and battery bank to power multiple lamps.

Solar LED lantern.

Rural

In rural India, solar lamps, commonly called solar lanterns, using either LEDs or CFLs, are being used to replace kerosene lamps, candles and other cheap alternatives of lighting. Especially in areas where electricity is otherwise difficult to access, solar lamps are very useful and it also improves the quality of life.

SOLAR FURNACE

A solar furnace is any device that creates heat by concentrating sunlight through the use of reflectors. There are many different sizes of solar furnaces that you can purchase or build yourself. In addition, you will find that there is a wide array of uses for these kinds of furnaces.

A small furnace can be used for cooking, without having to use any firewood. A larger furnace can generate electricity for powering additional items within a home - even powering the entire home! A homeowner could even retrofit a new furnace into the existing home.

While this concept has worked to introduce the technology to developing countries, it still has only limited success. The hope is that introducing it to developing countries will minimize firewood-related deforestation, which is an ongoing concern with the environment.

So how exactly does a solar furnace work? The principle of operation consists of two mirrors pointing at a focus. The intensity of sunlight in that focus will increase approximately three times. The reason for this is because instead of only getting light from the sun, the focus gets light from the sun and two mirrors.

Two mirrors pointing at a focus is not going to generate enough solar energy for you to do much. Once you have 10 mirrors pointing at a focus, you will begin to get closer to levels that are capable of heating and cooking food.

The main limitation you are going to find with how much solar energy you can get to focus on one spot has to do with how accurately the mirrors can be aimed at the focus. The solar furnace is something that is extremely useful and is quickly gaining in popularity despite this limitation.

If you are into building and constructing things on your own, you can certainly build your own solar furnace depending on what you are looking to accomplish. If you are not quite as ambitious, you can purchase them as well. Two great things about solar furnaces are that they are extremely versatile and they have no negative effects on the environment.

SOLAR POWER TOWER

Solar power towers convert sunshine into clean electricity. The technology uses many large, sun-tracking mirrors commonly referred to as heliostats to focus sunlight on a receiver at the top of a tower.

A heat transfer fluid heated in the receiver is used to generate steam, which, in turn, is used in a conventional turbine-generator to produce electricity. Early power towers such as the Solar One plant used steam as the heat transfer fluid. Current power towers, based on Solar Two, use molten nitrate salt because of its superior heat transfer and energy storage capabilities.

Solar One: The First Generation of Power Tower Plant

Solar One was the world's largest power tower plant, which operated from 1982 to 1988 in the Mojave Desert. The Solar One thermal storage system worked by storing heat in the form of steam generated using solar energy in a tank filled with rocks and sand and using oil as the heat-transfer fluid.

The Solar One thermal storage system extended the power generation capability of the plant into the night and provided heat for generating low-grade steam for keeping parts of the plant warm during off-hours and for morning start-up.

Unfortunately, the Solar One thermal storage system was complex and thermodynamically inefficient. Solar One also showed the disadvantages of a water/steam system, such as the intermittent operation of the turbine due to cloud transience and lack of effective thermal storage.

Solar Two: The Second Generation of Power Tower Plant

The expansion of Solar One to Solar Two required a new molten-salt heat transfer system and a new control system which allowed the heliostats to track the sun. This included the receiver, thermal storage, piping, and a steam generator. The Solar One heliostat field, the tower, and the turbine or generator required only minimal modifications.

The tower was composed of a series of panels, each made of 32 thin-walled, stainless steel tubes, through which the molten salt flows in a serpentine path. The panels form a cylindrical shell surrounding piping, structural supports, and control equipment.

A black Pyromark paint which is robust, resistant to high temperatures and thermal cycling, and absorbs 95% of the incident sunlight, was used to coat the external surfaces of the tubes. The receiver design was optimized to absorb a maximum amount of solar energy while reducing the heat losses due to convection and radiation.

The design included laser-welding, sophisticated tube-nozzle-header connections, a tube clip design that facilitated tube expansion and contraction, and non-contact flux measurement devices, which allowed the receiver to rapidly change temperature without being damaged.

Advantages of using Molten Salt

A variety of fluids were tested to transport the sun's heat, including water, air, oil, and sodium, before molten salt was chosen: it is liquid at atmospheric pressure, provides an efficient, low-cost medium in which to store thermal energy, its operating temperatures are compatible with today's high-pressure and high-temperature steam turbines, and it is non-flammable and nontoxic. In addition, molten salt is used in the chemical and metals industries as a heat-transport fluid, so experience with molten-salt systems exists for non-solar applications.

The Salt Mixture

A mixture of 60 percent sodium nitrate and 40 percent potassium nitrate was employed as the salt storage medium and is still used in modern solar power towers. This salt melts at 220 °C and is maintained in a molten state of 290 °C in the 'cold' storage tank. It then travels through the receiver where it is heated to 565 °C and then on to a 'hot' tank for storage.

Hot salt is pumped to a steam generating system when power is needed from the plant. These hot salts produce superheated steam for a conventional Rankine-cycle turbine generator system. From the steam generator, the salt is returned to the cold tank where it is stored and eventually reheated in the receiver.

Metal Corrosion in the Molten-salt Environment

All pipes, valves, and vessels for hot salt were constructed from stainless steel because of its corrosion resistance in the molten-salt environment, while the cold-salt system is made from mild carbon steel.

Benefits of Solar Power Towers

Like all solar technologies, solar power towers are fueled by sunshine and do not release greenhouse gases. Solar power towers are unique among solar electric technologies in their ability to efficiently store solar energy and dispatch electricity to the grid when needed, even at night or during cloudy weather.

Despite the central tower being torn down in 2009, the success of Solar One and Two in the US prompted many other countries to develop their own installations, including Spain and Israel.

Environmental Impact of Solar Power Towers

No hazardous gaseous or liquid emissions are released during operation of the solar power tower plant. If a salt spill occurs, the salt will freeze before significant contamination of the soil occurs. Salt is picked up with a shovel and can be recycled if necessary. If the power tower is hybridized with a conventional fossil plant, emissions will be released from the non-solar portion of the plant.

Solar power towers also reduce a country's reliance on fossil fuels; in Israel for example, Ashalim will prevent 110,000 tons per year of CO_2 emissions, being released by burning fossil fuels.

There is evidence, however, that such large installations that concentrate solar rays could be detrimental to bird populations, which can experience scorched feathers, or even death when close by. In order to avoid such tragedies, a maximum of four mirrors are focused on any one spot during standyby.

SOLAR AIR CONDITIONING

Solar air conditioning refers to any air conditioning (cooling) system that uses solar power.

This can be done through passive solar, solar thermal energy conversion and photovoltaic conversion (sunlight to electricity). The U.S. Energy Independence and Security Act of 2007 created 2008 through 2012 funding for a new solar air conditioning research and development program, which should develop and demonstrate multiple new technology innovations and mass production economies of scale.

Photovoltaic (PV) Solar Cooling

Photovoltaics can provide the power for any type of electrically powered cooling be it conventional compressor-based or adsorption/absorption-based, though the most common implementation is with compressors. For small residential and small commercial cooling (less than 5 MWh/a) PV-powered cooling has been the most frequently implemented solar cooling technology. The reason for this is debated, but commonly suggested reasons include incentive structuring, lack of residential-sized equipment for other solar-cooling technologies, the advent of more efficient electrical coolers, or ease of installation compared to other solar-cooling technologies (like radiant cooling).

Since PV cooling's cost effectiveness depends largely on the cooling equipment and given the poor efficiencies in electrical cooling methods until recently it has not been cost effective without subsidies. Using more efficient electrical cooling methods and allowing longer payback schedules is changing that scenario.

For example, a 100,000 BTU U.S. Energy Star rated air conditioner with a high seasonal energy efficiency ratio (SEER) of 14 requires around 7 kW of electric power for full cooling output on a hot day. This would require over a 20 kW solar photovoltaic electricity generation system with storage.

A solar-tracking 7 kW photovoltaic system would probably have an installed price well over $20,000 USD (with PV equipment prices currently falling at roughly 17% per year). Infrastructure, wiring, mounting, and NEC code costs may add up to an additional cost; for instance a 3120 watt solar panel grid tie system has a panel cost of $0.99/watt peak, but still costs ~$2.2/watt hour peak. Other systems of different capacity cost even more, let alone battery backup systems, which cost even more.

A more efficient air conditioning system would require a smaller, less-expensive photovoltaic system. A high-quality geothermal heat pump installation can have a SEER in the range of 20 (±). A 100,000 BTU SEER 20 air conditioner would require less than 5 kW while operating.

Newer and lower power technology including reverse inverter DC heat pumps can achieve SEER ratings up to 26.

There are new non-compressor-based electrical air conditioning systems with a SEER above 20 coming on the market. New versions of phase-change indirect evaporative coolers use nothing but a fan and a supply of water to cool buildings without adding extra interior humidity (such as at McCarran Airport Las Vegas Nevada). In dry arid climates with relative humidity below 45% (about 40% of the continental U.S.) indirect evaporative coolers can achieve a SEER above 20, and up to SEER 40. A 100,000 BTU indirect evaporative cooler would only need enough photovoltaic power for the circulation fan (plus a water supply).

A less-expensive partial-power photovoltaic system can reduce (but not eliminate) the monthly amount of electricity purchased from the power grid for air conditioning (and other uses). With American state government subsidies of $2.50 to $5.00 USD per photovoltaic watt, the amortized cost of PV-generated electricity can be below $0.15 per kWh. This is currently cost effective in some areas where power company electricity is now $0.15 or more. Excess PV power generated when air conditioning is not required can be sold to the power grid in many locations, which can reduce (or eliminate) annual net electricity purchase requirement.

Superior energy efficiency can be designed into new construction (or retrofitted to existing buildings). Since the U.S. Department of Energy was created in 1977, their Weatherization Assistance Program has reduced heating-and-cooling load on 5.5 million low-income affordable homes an average of 31%. A hundred million American buildings still need improved weatherization. Careless conventional construction practices are still producing inefficient new buildings that need weatherization when they are first occupied.

It is fairly simple to reduce the heating-and-cooling requirement for new construction by one half. This can often be done at no additional net cost, since there are cost savings for smaller air conditioning systems and other benefits.

Geothermal Cooling

Earth sheltering or Earth cooling tubes can take advantage of the ambient temperature of the Earth to reduce or eliminate conventional air conditioning requirements. In many climates where the majority of humans live, they can greatly reduce the buildup of undesirable summer heat, and also help remove heat from the interior of the building. They increase construction cost, but reduce or eliminate the cost of conventional air conditioning equipment.

Earth cooling tubes are not cost effective in hot humid tropical environments where the ambient Earth temperature approaches human temperature comfort zone. A solar chimney or photovoltaic-powered fan can be used to exhaust undesired heat and draw in cooler, dehumidified air that has passed by ambient Earth temperature surfaces. Control of humidity and condensation are important design issues.

A geothermal heat pump uses ambient Earth temperature to improve SEER for heat and cooling. A deep well recirculates water to extract ambient Earth temperature (typically at 2 gallons of water per ton per minute). These "open loop" systems were the most common in early systems, however water quality could cause damage to the coils in the heat pump and shorten the life of the equipment. Another method is a closed loop system, in which a loop of tubing is run down a well or wells, or in trenches in the lawn, to cool an intermediate fluid. When wells are used, they are backfilled with Bentonite or another grout material to ensure good thermal conductivity to the earth.

In the past the fluid of choice was a 50/50 mixture of propylene glycol because it is non-toxic unlike ethylene glycol (which is used in car radiators). Propylene glycol is viscous, and would eventually gum up some parts in the loop(s), so it has fallen out of favor. Today, the most common transfer agent is a mixture of water and ethyl alcohol (ethanol).

Ambient earth temperature is much lower than peak summer air temperature, and much higher than the lowest extreme winter air temperature. Water is 25 times more thermally conductive than air, so it is much more efficient than an outside air heat pump, (which becomes less effective when the outside temperature drops in Winter).

The same type of geothermal well can be used without a heat pump but with greatly diminished results. Ambient Earth temperature water is pumped through a shrouded radiator (like an automobile radiator). Air is blown across the radiator, which cools without a compressor-based air conditioner. Photovoltaic solar electric panels produce electricity for the water pump and fan, eliminating conventional air-conditioning utility bills. This concept is cost-effective, as long as the location has ambient Earth temperature below the human thermal comfort zone (not the tropics).

Solar Open-loop Air Conditioning using Desiccants

Air can be passed over common, solid desiccants (like silica gel or zeolite) or liquid desiccants (like lithium bromide/chloride) to draw moisture from the air to allow an efficient mechanical or evaporative cooling cycle. The desiccant is then regenerated by using solar thermal energy to dehumidfy, in a cost-effective, low-energy-consumption, continuously repeating cycle. A photovoltaic system can power a low-energy air circulation fan, and a motor to slowly rotate a large disk filled with desiccant.

Energy recovery ventilation systems provide a controlled way of ventilating a home while minimizing energy loss. Air is passed through an "enthalpy wheel" (often using silica gel) to reduce the cost of heating ventilated air in the winter by transferring heat from the warm inside air being exhausted to the fresh (but cold) supply air. In the summer, the inside air cools the warmer incoming supply air to reduce ventilation cooling costs. This low-energy fan-and-motor ventilation system can be cost-effectively powered by photovoltaics, with enhanced natural convection exhaust up a solar chimney - the downward incoming air flow would be forced convection (advection).

A desiccant like calcium chloride can be mixed with water to create an attractive recirculating waterfall, that dehumidifies a room using solar thermal energy to regenerate the liquid, and a PV-powered low-rate water pump.

Active solar cooling wherein solar thermal collectors provide input energy for a desiccant cooling system. There are several commercially available systems that blow air through a desiccant impregnated medium for both the dehumidification and the regeneration cycle. The solar heat is one way that the regeneration cycle is powered. In theory packed towers can be used to form a counter-current flow of the air and the liquid desiccant but are not normally employed in commercially available machines. Preheating of the air is shown to greatly enhance desiccant regeneration. The packed column yields good results as a dehumidifier/regenerator, provided pressure drop can be reduced with the use of suitable packing.

Passive Solar Cooling

In this type of cooling solar thermal energy is not used directly to create a cold environment or drive any direct cooling processes. Instead, solar building design aims at slowing the rate of heat transfer into a building in the summer, and improving the removal of unwanted heat. It involves a good understanding of the mechanisms of heat transfer: heat conduction, convective heat transfer, and thermal radiation, the latter primarily from the sun.

For example, a sign of poor thermal design is an attic that gets hotter in summer than the peak outside air temperature. This can be significantly reduced or eliminated with a cool roof or a green roof, which can reduce the roof surface temperature by 70 °F (40 °C) in summer. A radiant barrier and an air gap below the roof will block about 97% of downward radiation from roof cladding heated by the sun.

Passive solar cooling is much easier to achieve in new construction than by adapting existing buildings. There are many design specifics involved in passive solar cooling. It is a primary element of designing a zero energy building in a hot climate.

Solar Closed-loop Absorption Cooling

The following are common technologies in use for solar thermal closed-loop air conditioning.

- Absorption: NH_3/H_2O or Ammonia/Water.
- Absorption: Water/Lithium Bromide.
- Absorption: Water/Lithium Chloride.
- Adsorption: Water/Silica Gel or Water/Zeolite.
- Adsorption: Methanol/Activated Carbon.

Active solar cooling uses solar thermal collectors to provide solar energy to thermally driven chillers (usually adsorption or absorption chillers). Solar energy heats a fluid that provides heat to the generator of an absorption chiller and is recirculated back to the collectors. The heat provided to the generator drives a cooling cycle that produces chilled water. The chilled water produced is used for large commercial and industrial cooling.

Solar thermal energy can be used to efficiently cool in the summer, and also heat domestic hot water and buildings in the winter. Single, double or triple iterative absorption cooling cycles are used in different solar-thermal-cooling system designs. The more cycles, the more efficient they are. Absorption chillers operate with less noise and vibration than compressor-based chillers, but their capital costs are relatively high.

Efficient absorption chillers nominally require water of at least 190 °F (88 °C). Common, inexpensive flat-plate solar thermal collectors only produce about 160 °F (71 °C) water. High temperature flat plate, concentrating (CSP) or evacuated tube collectors are needed to produce the higher temperature transfer fluids required. In large scale installations there are several projects successful both technical and economical in operation worldwide including, for example, at the headquarters of Caixa Geral de Depósitos in Lisbon with 1,579 square metres (17,000 sq ft) solar collectors and 545 kW cooling power or on the Olympic Sailing Village in Qingdao/China. In 2011 the most powerful plant at Singapore's new constructed United World College will be commissioned (1500 kW).

These projects have shown that flat plate solar collectors specially developed for temperatures over 200 °F (93 °C) (featuring double glazing, increased backside insulation, etc.) can be effective and cost efficient. Where water can be heated well above 190 °F (88 °C), it can be stored and used when the sun is not shining.

The Audubon Environmental Center at the Ernest E. Debs Regional Park in Los Angeles has an example solar air conditioning installation, which failed fairly soon after commissioning and is no longer being maintained. The Southern California Gas Co. (The Gas Company) is also testing the practicality of solar thermal cooling systems at their Energy Resource Center (ERC) in Downey, California. Solar Collectors from Sopogy and Cogenra were installed on the rooftop at the ERC and are producing cooling for the building's air conditioning system. Masdar City in the United Arab Emirates is also testing a double-effect absorption cooling plant using Sopogy parabolic trough collectors, Mirroxx Fresnel array and TVP Solar high-vacuum solar thermal panels.

For 150 years, absorption chillers have been used to make ice (before the electric light bulbs were invented). This ice can be stored and used as an "ice battery" for cooling when the sun is not shining, as it was in the 1995 Hotel New Otani Tokyo in Japan. Mathematical models are available in the public domain for ice-based thermal energy storage performance calculations.

The ISAAC Solar Icemaker is an intermittent solar ammonia-water absorption cycle. The ISAAC uses a parabolic trough solar collector and a compact and efficient design to produce ice with no fuel or electric input, and with no moving parts.

Providers of solar cooling systems include ChillSolar, SOLID, Sopogy, Cogenra, Industrial Solar and TVP Solar for commercial installations and ClimateWell, Fagor-Rotartica, SorTech and Daikin mostly for residential systems. Cogenra uses solar co-generation to produce both thermal and electric energy that can be used for cooling.

Solar Cooling Systems Utilizing Concentrating Collectors

The main reasons for employing concentrating collectors in solar cooling systems are: high efficient air-conditioning through coupling with double/triple effect chillers; and solar refrigeration serving industrial end-users, possibly in combination with process heat and steam.

Concerning industrial applications, several studies in the recent years highlighted that there is a high potential for refrigeration (temperatures below 0 °C) in different areas of the globe (e.g., the Mediterranean, Central America). However, this can be achieved by ammonia/ water absorption chillers requiring high temperature heat input at the generator, in a range (120 ÷ 180 °C) which can only be satisfied by concentrating solar collectors. Moreover, several industrial applications require both cooling and steam for processes, and concentrating solar collectors can be very advantageous in the sense that their use is maximized.

SOLAR INVERTER

A solar inverter is a common inverter but uses energy from the sun that is solar energy. This type of inverter aids in changing the DC into AC using solar power. DC is the power which flows in one direction in the circuit and helps in providing current when there is no electricity. Direct current is used for small appliances such as gadgets, iPod, MP3 players, etc. where there is power stored in the battery. In the case of AC (alternative current) is the power that supplies back and forth inside the circuit. Generally, the AC power is used for household appliances. A solar inverter helps many devices that run on DC power to run on AC power so that the operator makes use of the AC power.

If you are thinking why to use this inverter as an alternative to the normal electric one, then it is because the solar inverter makes use of the solar energy which is obtainable in ample from the Sun and is clean and also pollution free.

The Solar Inverter is an essential device in any solar power system. Its basic function of the inverter is to change the variable Direct Current output of the solar panels into Alternating Current. The various electrical and electronic components connected in the circuit help in the conversion.

Solar Inverter.

The converted Alternating Current power is used for running your appliances like the TV, Refrigerator, Microwave, etc. For some particular applications, we can directly use the Direct Current power from the solar panel such as LED night lights, a cell phone charger. Generally, the power of a home solar power system is used for power AC loads.

Types of Solar Inverters

There are abundant of well-known creators of Solar Inverters all over the world. But there are a few types of solar inverters available in the market which includes the following.

Off Grid Inverters

Off grid inverters are used in remote systems wherein the solar inverter is fed DC power from a battery panel. This battery panel is charged by solar panels. Several such inverters have integrated with basic battery chargers which can be used to boost the battery from an AC power source.

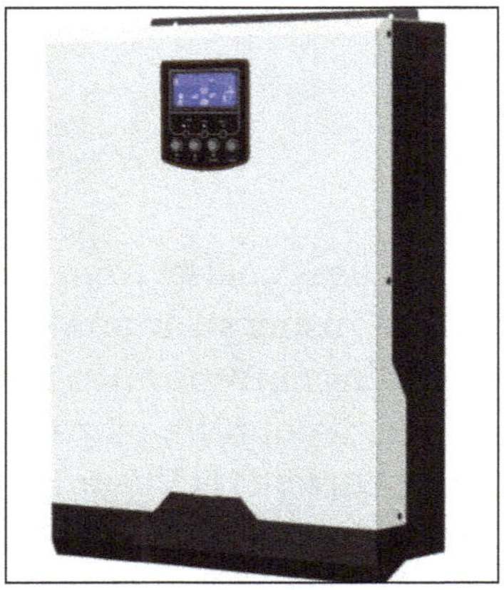

Off Grid Inverter.

Grid Tie Inverters

An inverter which is associated with the grid can be said to be a grid tie inverter. These inverter feed power in the electricity grid by corresponding phase and frequency. And, the frequency ranges 50Hz in India and 60Hz in North America of the o/p AC power with the effectiveness supplied AC power. These inverters are designed to automatically shut down once sensing a loss of supply from the utility.

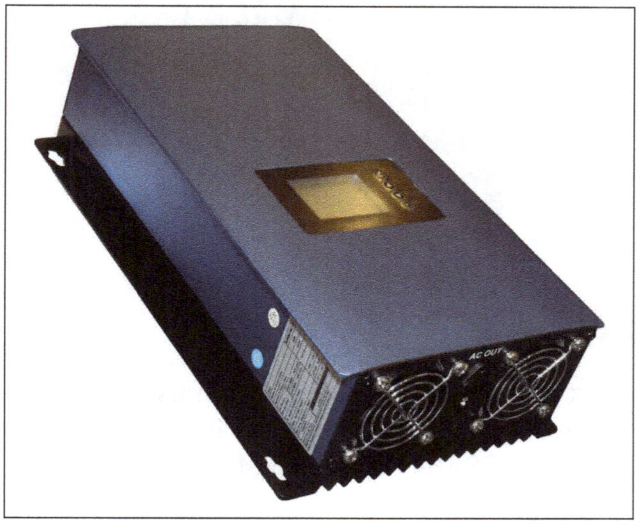

Grid Tie Inverter.

Battery Backup Inverters

These special types of inverters are particularly designed to draw energy from a battery. The charge of the battery is preserved by using an on-board charger and an additional energy is transferred to the grid. Such inverters have the ability to provide AC power to particular loads during power outages. They also have the anti-islanding function.

Battery backup Inverter.

Micro Inverters

Microinverters are modern in the solar industry. They are small, compact size and portable with a lot of performance. They include all the features of any central inverters.

Micro Inverter.

Solar Inverter and its Working

The main goal of this project is to design a system which uses solar energy for household loads using an inverter. The hardware requirements mainly include Bridge-MOSFETs drive, Step up Transformer, Voltage regulator, MOSFET driver, PWM inverter IC, Solar Panel, and Battery.

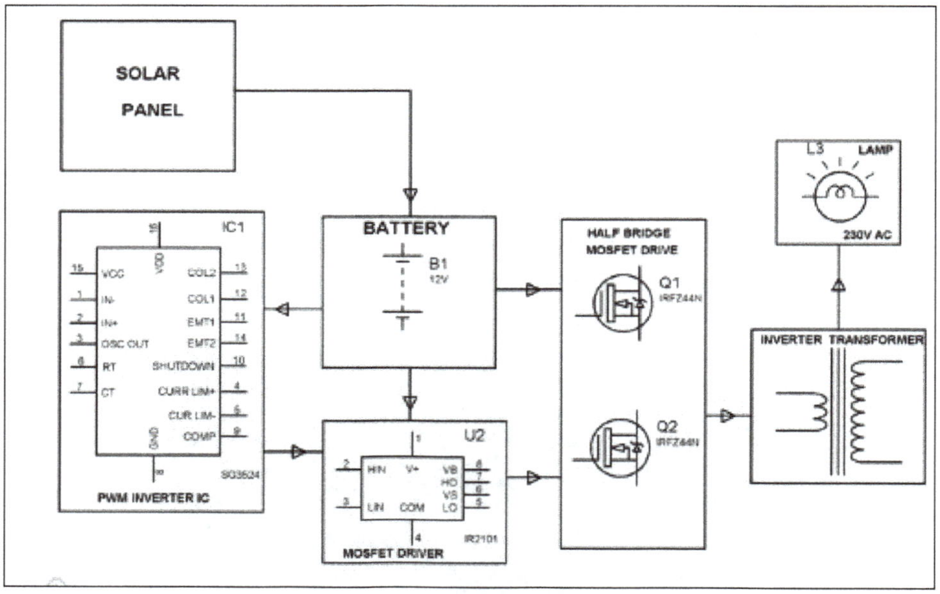

Solar Inverter Block Diagram.

Solar energy is changed to electrical energy with the help of photovoltaic cells. This energy is stored in batteries during daytime for the operation purpose whenever needed. The proposed system is designed to utilize solar energy for home loads using an inverter.

A solar inverter converts the DC (Direct Current) output of a PV solar panel into a utility frequency AC (alternating current) that can be fed into a commercial electrical grid (or) used by a local, off-line electrical n/w.

In this proposed system, the solar energy is stored in the battery from PV cells. This energy of the battery is altered to AC supply of 50Hz frequency using PWM inverter IC with MOSFET to the driver MOSFET bridge & step-up the voltage by a transformer, all in off-line n/w only but not grid tie type.

Solar Inverter Project Kit.

For sample purposes a battery of the SMF kind of and 5 AH (not supplied, it is generally used in small UPS) is necessary to be used as solar cells required would be of very high power type casting very high. Moreover, this project can be added to a charge controller for over voltage, under voltage protection and overload protection.

Advantages of Solar Inverter

After knowing in detail what is a solar inverter and how it is a suitable making appliance work at residential & industrial levels we must discuss about the several advantages of the solar inverter.

- Solar energy has constantly helped in decreasing the greenhouse effect and global warming.
- By using of solar based devices will help in saving money and also energy. Because many people have started using these devices.
- A solar inverter helps in changing the DC into batteries or AC. This supports people who use a partial amount of electricity.
- The synchronous solar inverter that helps small homeowners and also power companies as they are huge in size.
- The multifunction solar inverter is the finest among all and works powerfully. It converts the DC to AC very carefully which is suitable for commercial establishments.
- This inverter is cost effective, i.e. low cost than generators.
- Apart from these there are additional devices too, that make use of solar energy such as, solar heater, cooker.

Disadvantages of Solar Inverter

- Primarily, we need to shell out a lot of money for purchasing a solar inverter.
- It will work efficiently and produce DC only when the daylight is strong.
- The solar panels are used to attract the sunshine needs lots of space.
- Solar Inverters can work when there is no Sunshine but the battery which is available in that is charged fully with the help of Sunshine.

SOLAR CHIMNEY

A solar chimney is a type of passive solar heating and cooling system that can be used to regulate the temperature of a building as well as providing ventilation. Like a Trombe wall or solar wall, solar chimneys are a way to achieve energy efficient building design. Essentially, solar chimneys are hollow containers that connect the inside part of the building to the outside part of the building.

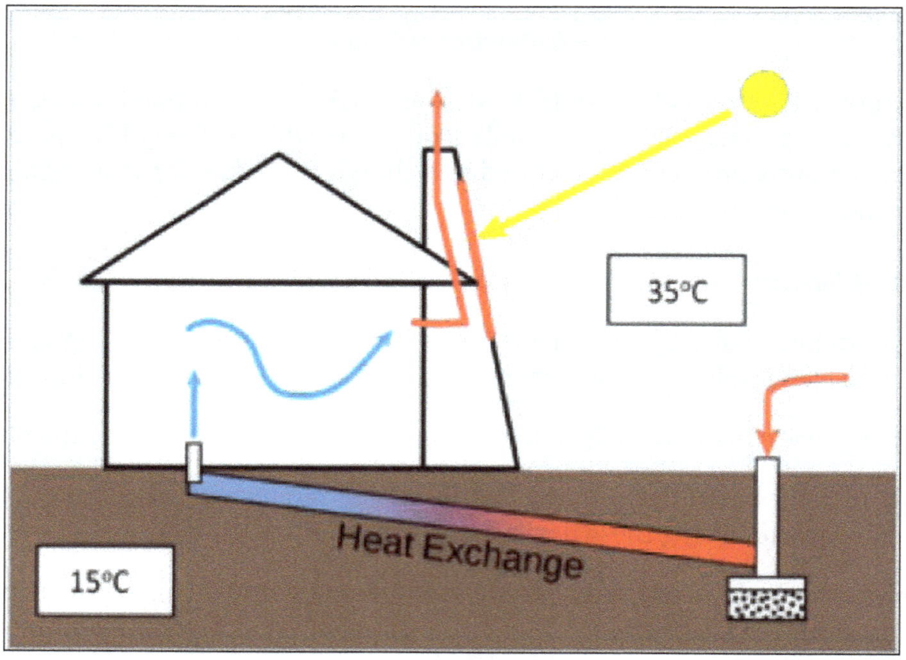

A diagram of a solar chimney that uses a secondary vent that travels below ground. This helps to cool intake air when the chimney is used for cooling.

How it Works

Solar chimneys are easy and inexpensive means to heat and ventilate a building. First, a chimney is built and coated in some dark or black material. It is coloured black because this minimizes the amount of sunlight that is reflected off of the chimney, absorbing more of the heat and ensuring more of the heat is transferred to the air inside the building. As well, these chimneys are generally placed on a South facing wall if the home is in the Northern hemisphere.

The process of heating a space using a solar chimney is fairly simple. When the solar radiation hits the side of the chimney, the column of air inside the chimney is heated. If the top exterior vents of the chimney are closed, the heated air is forced back into the living space. This provides a type of convective air heating. As the air cools in the room it is pulled back into the solar chimney, heating once again. When solar chimneys are used for heating, they operate similarly to Trombe walls.

Cooling a space using a solar chimney is slightly different than cooling using a Trombe wall. Since a roof overhand cannot be installed in addition to a solar chimney, two additional vents are present. The first vent has been mentioned, the one at the top of the chimney. The second is at the opposite end of the building, providing an opening between the building and outside air to allow for ventilation. When solar radiation hits the side of the chimney, the column of air inside the chimney is again heated. The vent at the top of the chimney is kept open so this heated air is not trapped. This heated air is pulled up and out of the chimney, pulling new air in from the outside and creating a sort of "draft" that provides cool, fresh air into the building.

SOLAR CHARGER

A solar charger is a charger that employs solar energy to supply electricity to devices or batteries. They are generally portable.

Suntactics sCharger-5.

Solar chargers can charge lead acid or Ni-Cd battery banks up to 48 V and hundreds of ampere-hours (up to 4000 Ah) capacity. Such type of solar charger setups generally use an intelligent charge controller. A series of solar cells are installed in a stationary location (ie: rooftops of homes, base-station locations on the ground etc.) and can be connected to a battery bank to store energy for off-peak usage. They can also be used in addition to mains-supply chargers for energy saving during the daytime.

Most portable chargers can obtain energy from the sun only. Some, including the Kinesis K3, and

GeNNex Solar Cell 2 can work either way (recharged by the sun or plugged into a wall plug to charge up). Examples of solar chargers in popular use include:

- Small portable models designed to charge a range of different mobile phones, cell phones, iPods or other portable audio equipment.
- Fold out models designed to sit on the dashboard of an automobile and plug into the cigar/12v lighter socket to keep the battery topped up while the vehicle is not in use.
- Flashlights/torches, often combined with a secondary means of charging, such as a kinetic (hand crank generator) charging system.
- Public solar chargers permanently installed in public places, such as parks, squares and streets, which anyone can use for free.

Front and back views of a small, portable solar charger with two AAA Batteries and USB output.

Voltage Regulator

A solar panel can produce a range of charging voltages depending upon sunlight intensity, so a voltage regulator must be included in the charging circuit so as to not over-charge (overvoltage) a device such as a 12 volt car battery.

Solar Chargers on the Market

Portable solar chargers are used to charge cell phones and other small electronic devices on the go. Chargers on the market today use various types of solar panels, ranging from thin film panels with efficiencies from 7-15% (amorphous silicon around 7%, CIGS closer to 15%), to the slightly more efficient monocrystalline panels which offer efficiencies up to 18%.

Solar charger integrated into a backpack.

The other type of portable solar chargers are those with wheels which enable them to be transported from one place to another and be used by a lot of people. They are semi-public, considering the fact that are used publicly but not permanently installed. A good example of this kind of portable solar charger is the Strawberry Mini device.

Creative Edge Solar-5, 1.2 w solar panel with 5000 mAh internal battery.

The solar charger industry has been plagued by companies mass-producing low efficiency solar chargers that don't meet the consumer's expectations. This in turn has made it hard for new solar charger companies to gain the trust of consumers. Solar companies are starting to offer high-efficiency solar chargers. When it comes to permanently installed public solar chargers, Strawberry energy Company from Serbia has invented and developed the first public solar charger for mobile devices, Strawberry Tree. Due to a built in rechargeable battery which stores energy, it can function without sunshine or at night. Other companies such as Voltaic Systems, Poweradd and others have started to push better products onto the market as well.

Public solar charger for mobile phones.

The JOOS Orange by Solar JOOS.

Rural African villagers holding portable solar charger.

Portable solar power is being utilized in developing countries to power lighting as opposed to utilizing kerosene lamps which are responsible for respiratory infections, lung and throat cancers, serious eye infections, cataracts as well as low birth weights. Solar power provides an opportunity for rural areas to "leapfrog" traditional grid infrastructure and move directly to distributed energy solutions.

Some solar chargers also have an on-board battery which is charged by the solar panel when not charging anything else. This allows the user to be able to use the solar energy stored in the battery to charge their electronic devices at night or when indoors.

Solar chargers can also be rollable or flexible and are manufactured using thin film PV technology. Rollable solar chargers may include Li-ion batteries.

Currently, foldable solar panels are coming down in price to the point that almost anyone can deploy one while at the beach, biking, hiking, or at any outdoor location and charge their cellphone, tablet, computer etc. As advances in the technology continue to be made, new products will be able to be expand into third world countries faster than ever before. Recently, billionaire Elon Musk unveiled a new battery system that may allow off-grid systems that rely solely on solar power to recharge them to be deployed anywhere on the planet, thus possibly ending the need for strictly grid-based energy systems.

SOLAR COOKERS

Solar cooker, also called solar oven, is a device that harnesses sunlight as a source of heat for cooking foodstuffs. The solar oven is a simple, portable, economical, and efficient tool.

Types of Solar Cookers

- For household cooking: Box type solar cooker.
- For community cooking: Concentrator type solar cookers also called as Scheffler's model.

Box Type Solar Cooker

It is like a hot box, in which food can be cooked. It is the most commonly used model for household cooking.

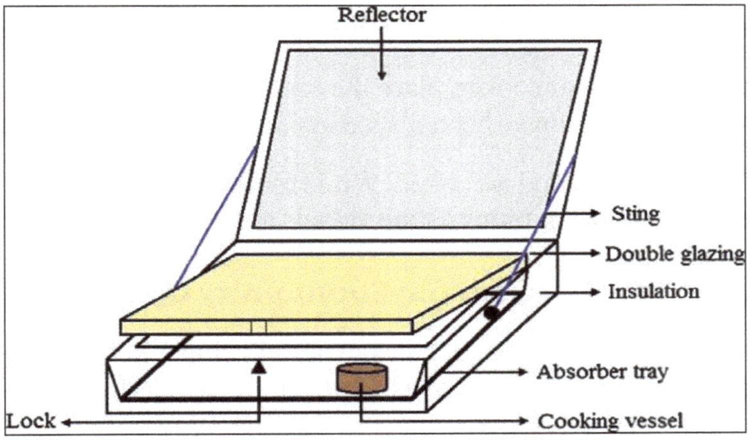

Important parts of a box type solar cooker.

- Outer Box: The outer box of a solar cooker is generally made of G.I. or aluminum sheet or fibre reinforced plastic.
- Inner Cooking Box (Tray): This is made from aluminum sheet. The inner cooking box is slightly smaller than the outer box. It is coated with black paint so as to easily absorb solar radiation and transfer the heat to the cooking pots.
- Double Glass Lid: A double glass lid covers the inner box or tray. This cover is slightly larger than the inner box. The two glass sheets are fixed in an aluminum frame with a spacing of 2 centimeters between the two glasses. This space contains air which insulates and prevents heat escaping from inside. A rubber strip is affixed on the edges of the frame to prevent any heat leakage.
- Thermal Insulator: The space between the outer box and inner tray including bottom of the tray is packed with insulating material such as glass wool pads to reduce heat losses from the cooker. This insulating material should be free from volatile materials.
- Mirror: Mirror is used in a solar cooker to increase the radiation input on the absorbing space and is fixed on the inner side of the main cover of the box. Sunlight falling on the

mirror gets reflected from it and enters into the tray through the double glass lid. This radiation is in addition to the radiation entering the box directly and helps to quicken the cooking process by raising the inside temperature of the cooker.

- Containers: The cooking containers (with cover) are generally made of aluminum or stainless steel. These vessels are also painted black on the outer surface so that they also absorb solar radiation directly.

Box type cookers with electrical back up are also available in the market.

Process of Cooking Food using a Box Type Solar Cooker

- Keep the solar cooker in the sun in open space free from any shadow. Keep the cooker in the sun at least for 45 minutes before loading it with cooking pots. This will keep the cooker ready for cooking and reduce the cooking time.
- Adjust the cooker in such a way that the reflecting mirror faces the sun and the reflected rays fall on the transparent glass lid. Tighten the position fixing hinges of the mirror in this position.
- Open the glass lid of the solar cooker, place the cooking pots inside it and close the lid properly. Once the cooking vessels have been placed inside the cooker the lid should not be opened.
- Open the lid fully once the food is cooked. While removing the cooking vessels after the food has been cooked, use cloth napkins as the vessels are very hot.

Concentrator Type Solar Cookers for Community Cooking

It works on the principle of solar energy concentration using a Reflecting Parabolic Solar Concentrator. A parabolic solar concentrator is used for concentrating solar radiation on a focal area where the cooking vessel is placed.

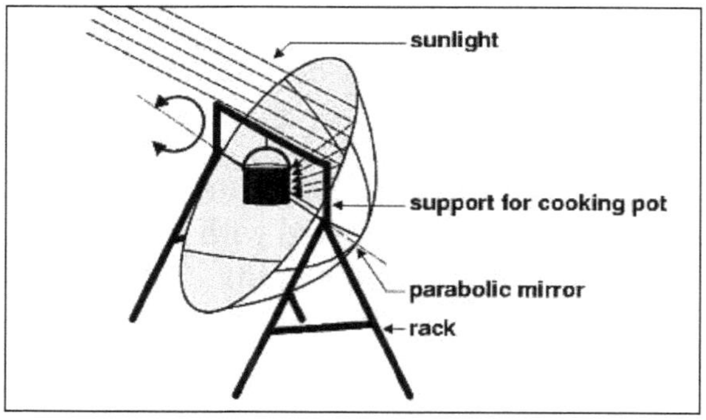

Parts of the Concentrator Type Solar Cooker

- Solar Concentrating Disc (Primary Reflector) - The disc which helps in concentrating solar energy to a focal point.
- Automatic Tracking System - With the help of a simple automatic mechanical tracking system the solar disc rotates in the direction of the movement of the Sun to give continuous and accurate solar energy concentration.

- Secondary Reflector - This is provided opening in the north-facing wall of the kitchen or the cooking place just below the cooking vessel. This reflector receives the concentrated solar radiation and reflects it on to the bottom of the cooking vessel.
- Cooking vessels.

Process of Cooking Food using a Concentrator Type Solar Cooker

- Installation:
  - The Solar disc is installed in the open shadow-free area or on terrace tops facing the South.
  - The cooking place and vessels faces the North (at the same ground level). The reflection of the disc falls on a secondary reflector housed in an opening in the North kitchen wall.
- Working of a community solar cooker:
  - Daily in the morning the disc is manually oriented so as to face the morning sun – in the east.
  - The daily orientation action winds the clock mechanism to work throughout the day and the disc starts rotating in the direction of the Sun guided by the Sundial.
  - The Cooker begins to work automatically as the concentrated solar energy is directed to the cooking vessel.
  - The primary concentrated reflection falls on the secondary reflector, which is placed right below the cooking vessel. The secondary reflector diverts the solar energy on the cooking vessel and the heating begin.
  - The seasonal adjustment of the disc is required to be done once in six months due to shift of the Sun's position, with respect to the Earth axis.

Make your own Solar Cooker

Simple things which are readily available around the house can be used to create a solar cooker. In your 'do it yourself' solar cooker you can cook a delicious meal without lighting a match.

Material Required

- Two reasonably large cardboard boxes - a size that can hold two large bowls with some room to spare. One box should fit into the other.
- Straw/Old newspapers – Collect and crush a few old newspapers, better still if you can get some straw. These will be used for insulation.
- A transparent glass cut to the dimensions of the top of the box – to be used as the lid of the inner box.
- Aluminium foil.

- Black acrylic paint.
- Packing tape.
- Scissors.
- Glass sheet.
- 2 aluminium bowls with metal lids.
- Half bowl rice; vegetable of choice and spices.

Methodology

- The inner Box: Clean out the box, tape up all the sides except the top. Ensure that the box is sturdy. With a pair of sharp scissors cut off the top four top flaps of the inner box. Tape up the rough edges. Now, paint the inside of the box a very very dark black - preferably using acrylic paint. Leave the box to dry.
- The outer box: Clean out the box, tape up all the sides except the top. Ensure that the box is strong. With the scissors cut out large pieces of aluminium foil - the kind that is used to wrap food. Paste the pieces on the inner side of the four top flaps. Leave the box to dry.
- The insulation: Use materials that are poor conductors of heat - rolled up newspaper, straw, husk etc. Place a layer of this material at the bottom of the big box and then place the smaller box on top. Next, fill out the gaps between the four sides of the boxes tightly with insulation material. Once this is done the two boxes should become one unit.
- The glass top: Place the glass top on the inner box.
- The meal: Select two non reflective bowls with lids. In one bowl put half a cup of washed rice and pour two cups of water. In the other bowl mix your favourite vegetable with a spoonful of oil, salt to taste, a pinch of turmeric and green chillies. Place both bowls carefully in your solar cooker and set it out in the sun. Wait for two hours, and the delicious food is ready, cooked only with the pure goodness of nature.

Advantages of a Solar Cooker

- No requirement of cooking gas or kerosene, electricity, coal or wood.
- No need to spend on fuel, as solar energy is available free.
- Food cooked in solar cooker is nutritious. About 10-20% of protein retention is more as compared to that in conventional cooking .Vitamin thiamine retention is about 20 to 30% more whereas vitamin A is retained 5 to 10% more when food is cooked in solar cooker.
- Solar cooking is pollution free and safe.
- Solar cookers come in various sizes. Based on the number of family members, the size of the cooker can be chosen.

- All cooking activities (like boiling, roasting) can be done using a solar cooker.
- There are government schemes which offer subsidies to purchase solar cookers.

Disadvantages of a Solar Cooker

- Adequate sunshine is required for cooking.
- Takes longer time to cook food than the conventional cooking methods.

SOLAR POWERED FAN

A solar fan is a mechanical fan powered by solar panels. The solar panels are either mounted on the device or are installed independently. Solar fans mostly do not require secondary power sources other than solar power, as most of them are used for cooling purposes during day time. Some types are also used for heating purposes. It runs the fastest when it is the hottest outside providing savings on air conditioning costs.

A solar fan with externally mounted solar panels

Applications

Attic heat is caused by the sun baking down on the roof surface. This sunlight causes the roofing material to heat up, thus transferring the heat into the attic via thermal transfer. The buildup of heat is generally dissipated by normal convection, i.e.: as the air is heated it expands and rises, flowing out of the attic through vents placed at or near the peak, while fresh air is pulled into the attic from intake vents that are generally in the eave (soffit) area. However, most roof vents are inadequate at expelling the heat and require the use of a powered vent fan to remove heat effectively. A solar powered attic fan is an effective way to remove attic heat using only the power of the sun to run the fan.

Solar-Powered Attic Fan - Roof Mounted

A typical roof-mounted solar attic fan will be effective in removing excessive heat from the attic. When mounted near the peak the solar fan forces hot air out while pulling fresh air in via eave/soffit vents. This exchange of air keeps the attic space from overheating. Solar attic fans not only lower the cooling costs in your home, they also extend the life of your shingle roof.

Solar-powered Attic Fan - Gable Mounted

It is similar to the roof mounted solar fan, except that its installation is a little different. Most attic fans are installed on the roof surface, whereas gable fans are installed in the gables, inside the attic and behind an existing louvered gable vent as the name suggests. They are mostly installed in places like greenhouse, cabin, RV's garages or barns and other storage places where heat and moisture are usually present. These types of fans are generally installed behind intake grills to provide ventilation. Gable Solar Fans are usually installed high on the gable close to the ridge and are also coupled with soffit or roof vents for well-balanced intake and exhaust air avenues. The airflow depends upon the level of sunlight when there is more sunlight, the airflow will be more resulting in enhanced cooling. It is also capable of lowering the air conditioning costs by cooling the attic.

Since the only source of energy is from the solar panel for a solar fan. It is important to have a larger solar panel to get the better performance from the solar fan. Amtrak Solar makes one of the most powerful solar fans in United States of America.

In 2007, solar fans were installed on the terminal roofs of Honolulu International Airport, in addition to the mansion of Indiana governor. In 2009 the Indiana State Senate passed a bill giving a tax deduction to homeowners with solar fans.

Advantages

- A solar fan is more environmentally friendly.
- Though initially the installation may be costlier, in the long term it would turn out to be cheaper because it does not use power from the utility grid and may provide savings up to 30 percent on air conditioning costs.
- Risk of electric accidents are eliminated as there are no electric cords attached to the solar fan.
- A cordless appliance offers much more mobility compared to conventional fans.

SOLAR DRYER

Solar dryers are devices that use solar energy to dry substances, especially food. There are two general types of solar dryers: Direct and indirect.

Direct

Direct solar dryers expose the substance to be dehydrated to direct sunlight. Historically, food and clothing was dried in the sun by using lines, or laying the items on rocks or on top of tents. In Mongolia cheese and meat are still traditionally dried using the top of the ger (tent) as a solar dryer. In these systems the solar drying is assisted by the movement of the air (wind) that removes the more saturated air away from the items being dried. More recently, complex drying racks and solar tents were constructed as solar dryers.

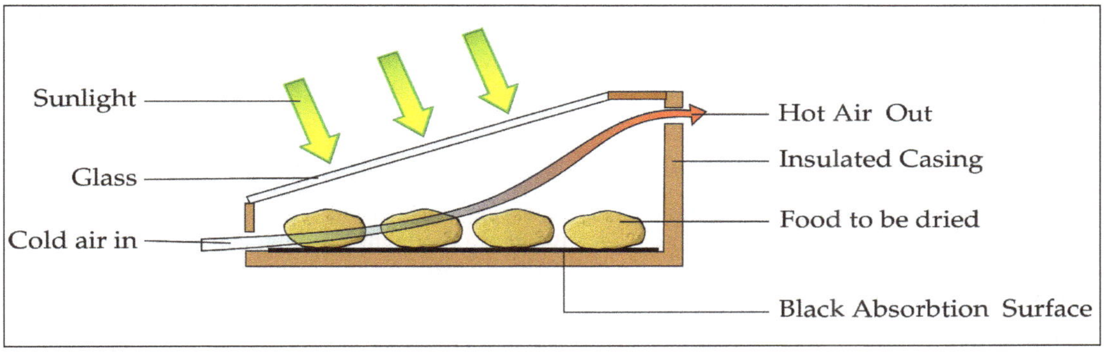

One modern type of solar dryer has a black absorbing surface which collects the light and converts it to heat; the substance to be dried is placed directly on this surface. These driers may have enclosures, glass covers and/or vents to in order to increase efficiency.

Indirect

Industrial indirect solar fruit and vegetable dryer.

In indirect solar dryers, the black surface heats incoming air, rather than directly heating the substance to be dried. This heated air is then passed over the substance to be dried and exits upwards often through a chimney, taking moisture released from the substance with it. They can be very simple, just a tilted cold frame with black cloth to an insulated brick building with active ventilation and a back-up heating system. One of the advantages of the indirect system is that it is easier to protect the food, or other substance, from contamination whether wind-blown or by birds, insects, or animals. Also, direct sun can chemically alter some foods making them less appetizing.

References

- How-a-solar-furnace-work: solarpoweristhefuture.com, Retrieved 19 August, 2019
- Bergesen, Joseph D.; Tähkämö, Leena; Gibon, Thomas; Suh, Sangwon (2016). "Potential Long-Term Global Environmental Implications of Efficient Light-Source Technologies". Journal of Industrial Ecology. 20 (2): 263. Doi:10.1111/jiec.12342

- How-solar-inverter-works-applications: edgefxkits.com, Retrieved 20 January, 2019
- Les Hamasaki. "10 Ton Solar Air Conditioning at the Debs Park Audubon Environmental Center in Los Angeles (6 minute video)". Retrieved 23 December 2007
- Solar-chimney, encyclopedia: energyeducation.ca, Retrieved 21 February, 2019
- Ayadi, Osama; Aprile, Marcello; Motta, Mario (1 January 2012). "Solar Cooling Systems Utilizing Concentrating Solar Collectors - An Overview". Energy Procedia. 30: 875–883. Doi:10.1016/j.egypro.2012.11.099. ISSN 1876-6102
- Solar-cooker, solar-energy, energy-production, energy: vikaspedia.in, Retrieved 22 March, 2019
- Heinz, Gunter & Hautzinger, Peter (2007). "Meat drying". Meat processing technology for small- to medium-scale producers. RAP Publication 2007/20. Bangkok, Thailand: Regional Office for Asia and the Pacific, Food and Agriculture Organization of the United Nations. ISBN 978-974-7946-99-4. Archived from the original on 23 May 2010
- Solar-water-heating: solar-energy-at-home.com, Retrieved 23 April, 2019
- Foster, Peter (2009-10-18). "$10 solar-powered lamp to help the poor". Telegraph.co.uk. Retrieved 2015-12-16
- Solar-pool-heating: solar-energy-at-home.com, Retrieved 24 May, 2019

Solar Power Applications in Spacecrafts

CHAPTER 6

Solar power is used to provide power to spacecraft which are operating in the inner solar system. Some of the applications of solar power in spacecraft are in solar thermal rockets, electric solar wind sails and solar electric propulsion. These diverse applications of solar power in spacecraft have been thoroughly discussed in this chapter.

SPACE SOLAR POWER

Space-based solar power is a method of using solar power satellites to collect solar energy so that it can be distributed for use all over the earth. With this amazing technology, space-based solar power is the future of power generation.

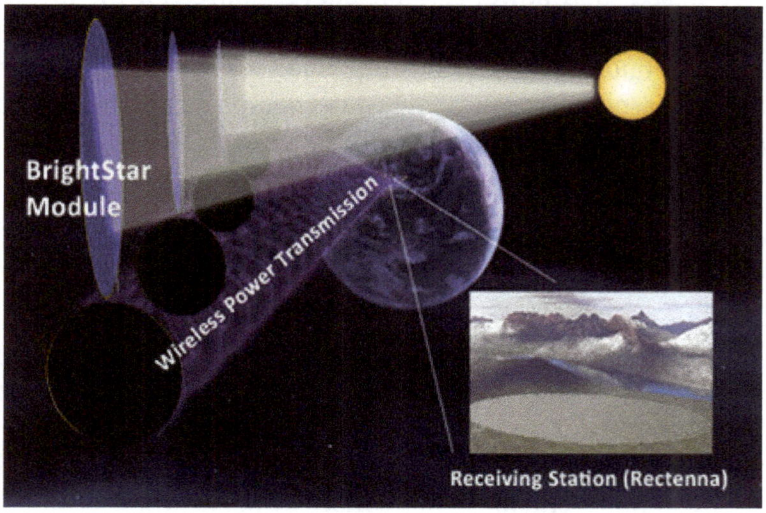

Solar Power Satellites (Powersats)

How does it work? Solar power satellites, otherwise known as powersats, orbit the earth and are designed to capture solar energy and transmit that energy to receiving stations that are situated thousands of miles from each other on the surface of the earth. These satellites are made up of a number of modules outfitted with light weight photovoltaic solar panels.

What is so great about this technology? It is clean, it is green, and it is safe. Collecting solar power in space is also more efficient than collecting solar power on the surface of the earth for many reasons.

Benefits of Space-based Solar Power

Space-based solar power is a method of collection that is not affected by weather on the surface, the seasons, or by the earth's atmosphere, which can act as a filter, reducing the strength of the solar energy. This means that it is far more reliable than surface solar panels and it is also a renewable source of energy.

What else makes space-based solar power a viable energy source for the future? It is fast. The energy can be transmitted from the solar power satellites to the receiving stations and then between receiving stations in just seconds. This is all accomplished via a wireless transfer of the energy and the result is dispatchable, baseload power, which can then power homes and businesses as effectively as the more traditional forms of power generation.

SOLAR POWER SATELLITE

Solar power satellites, otherwise known as powersats, orbit the earth and are designed to capture solar energy and transmit that energy to receiving stations that are situated thousands of miles from each other on the surface of the earth. These satellites are made up of a number of modules outfitted with light weight photovoltaic solar panels. It is clean, it is green, and it is safe. Collecting solar power in space is also more efficient than collecting solar power on the surface of the earth for many reasons.

The Solar Power Satellite (SPS) concept would place solar power plants in orbit above Earth, where they would convert sunlight to electricity and beam the power to ground-based receiving stations. The ground-based stations would be connected to today's regular electrical power lines that run to our homes, offices and factories here on Earth.

Why put solar power plants in space? The sun shines 24 hours a day in space, as if it were always noontime at the equator with no clouds and no atmosphere. Unlike solar power on the ground, the economy isn't vulnerable to cloudy days, and extra generating capacity and storage aren't needed for our nighttime needs. There is no variation of power supply during the course of the day and night, or from season to season. The latter problems have plagued ground based solar power concepts, but the SPS suffers none of the traditional limitations of ground-based solar power.

The satellites would be placed in so-called "geostationary" or "Earth synchronous" orbit, a 24-hour orbit which is thus synchronized with Earth's rotation, so that satellites placed there will stay stationary overhead from each's receiving antenna. (Likewise, today's communications satellites are put into geostationary orbit, and each TV satellite dish on the ground is pointed towards one satellite "stationary" in orbit.) The receiving antenna is called a "rectenna" (pronounced "rektenna").

Geostationary orbit is very high, 36,000 km (22,500 miles) above the surface of the Earth. It is far above the range of the Space Shuttle, which has a maximum range of about 1000 km (600 miles) above Earth's surface.

Aerospace company artists under contract to NASA and the US Dept. of Energy depicted the concept in the pictures at the beginning of this chapter. A technical depiction is given below.

The SPS will consist of a large sheet of solar cells mounted on a frame of steel-reinforced lunarcrete or astercrete. The solar cells produce electricity from sunlight with no moving parts. The only moving part on the satellite is the transmitter antenna(s) which slowly tracks the ground-based rectenna(s) while the solar cell array keeps facing the sun. Each transmitter antenna is connected to the solar cell array by two rotary joints with sliprings.

The transmitter on the SPS is an array of radio tubes (klystrons), waveguides, and heat radiators. They convert the electricity from the SPS solar cell power plant into a radio or microwave beam.

The ground-based rectenna consists of an array of antennas and standard electronics to convert the energy into regular AC electricity which can then be supplied into today's power lines.

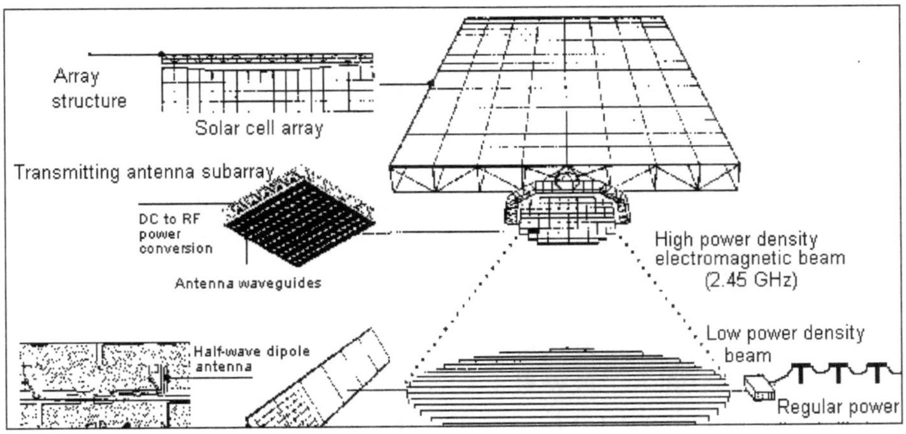

Basic concept of SPS.

An alternative concept, replacing the solar cell array with mirrors to heat a thermal engine to generate electricity, was called the "PowerSat". The word PowerSat seems smoother than the phrase Solar Power Satellite and especially its acronym SPS, and "PowerSat" is commonly used for both kinds of satellite, but sometimes there is confusion in esoteric professional circles because "PowerSat" was initially used for the alternative, thermal generation satellite. Hopefully, the generic sounding word "PowerSat" will start being used to denote either, as people could say "thermal PowerSat" to denote the thermal generator type.

The thermal PowerSat shared the same transmitting antenna and rectenna as the solar cell PowerSat.

The following photos show the testing of the SPS phased array transmitter-beam system, which took place across the California desert, the receiver a kilometer away from the transmitter.

Rectennas would probably be located above crops and livestock. A wire mesh screen can be placed between the rectenna and the ground which will cut out practically all the beam that leaks through the rectenna but will be over 80% transparent to light. In northern climates, adding glass to these structures and perhaps fluorescent lights could create greenhouses for a small additional cost (i.e., mechanized agriculture). Some designs have them floating offshore or on lakes, e.g., in Japan. (Having passed through Tokyo Narita airport, which is on a manmade island, I don't underestimate the economic viability of this concept.) They could double as fish farms for a relatively small add-on cost.

The SPS is not eclipsed by the Earth but is in sunlight continuously, 24 hours a day, due to its high orbit. The only exceptions are several days around the seasonal equinoxes, March 21 and September 22, when the satellite will be eclipsed briefly around midnite, for up to an hour and 12 minutes.

Notably, electricity will be needed for all sectors of space development, and thus the SPS can initially be used for in-space power in the early years of space development, with power being beamed to Earth after space-based industry grows. Thus, the SPS fits into a practical scale-up scenario. It may first be located at the site of major factories. Adding a transmitter will allow it to off-peak power to other sites, i.e., sell excess power - to facilities in nearby orbit, or the lunar surface, or to Earth economies. Electric powered orbit transfer vehicles could also use beamed power, replacing a heavy solar cell array with a lightweight rectenna to increase vehicle performance and reduce damages by transporting a solar cell array through the Van Allen radiation belt.

Earth based consumers have energy demands which vary between day and nite. Daytime power demand is typically twice nitetime power demand. "Peak" power is typically satisfied by power plants powered by natural gas and oil, whereas "baseload" is provided by coal, nuclear and hydroelectric power plants. If we want to replace fossil fuel and nuclear power, it may be possible to store power at the satellite at nitetime for beaming extra power in the daytime, e.g., by spinning up flywheels in the zero gravity of space (floating in zero gravity for practically no friction, thus very emergy efficient). Alternatively, we could scale the SPS for peak daily demand and at nite beam extra energy to other space based facilities equiped with energy storage devices for their own peak and varying consumption. The drop in power demand at nite could also be accomodated by shifting the consumption load of space based industry, e.g., operating the most energy intensive processes when it is nitetime below, as well as electric interorbital transport services. (If electric cars and buses take off in popularity on Earth, part of the nitetime shift could be in recharging of car and bus batteries, e.g., if rates are lower at nite.) The key is, of course, in the pricing structure. (The relative cost of SPS power is discussed in a moment.)

The solar power satellite fits into an asteroidal and lunar materials utilization scenario very well - it is composed of materials most abundant in asteroids near Earth and/or from the Moon, and it is made up of a small variety of simple parts mass produced in large quantity. The industry required in space to produce SPS components is relatively modest. Some design studies claim that more than 99% of an SPS can be made from asteroidal and/or lunar material.

The silicon solar cells can be made from lunar or asteroidal silicon, as silicon is the second most abundant element on the Moon and likewise in many kinds of asteroids. Purification of silicon is easier in the vacuum of space, and better crystals grow in zero gravity (due to no convection currents). The glass cover over the solar cells could be silica glass (silicon dioxide) - composed of the two most abundant elements on the Moon and likewise in many kinds of asteroids.

The SPS structure could be made from asteroidal nickel-iron steel or steel-reinforced lunarcrete or astercrete using cheap glass-ceramics or fiberglass composites. The waveguides could be made of glass ceramics. The vacuum tubes to generate the beam could be largely steel in terms of weight, with the small electrodes perhaps imported from Earth depending on the level of effort we put into processing the different kinds of asteroidal materials.

Given the advantages of manufacturing semiconductors in orbit, Silicon Valley could lose big business to competition from Silicon Orbit in the future. Computer chips are small and lightweight enough to bring back to Earth. A future generation of space-made chips may start to bring a close to many kinds of chip manufacturing on Earth, to the benefit of consumers and Earth's environment.

The first to go into business making silicon solar cells in space may well become the first to dominate that business. That may be judged by the first to experiment in orbit and get patents.

SPS Compared to other Power Sources

There are many different ways of generating electricity, e.g., coal fired power plants, nuclear energy, oil, natural gas and hydroelectric dams. The bottom line is that the sources that are used are those which are least expensive.

Likewise, there are many different ways we use energy, e.g., air conditioning, space heating, transportation, lighting and manufacturing. The best energy source is one which is flexible enough to be adapted for diverse uses.

Finally, our societies are requiring energy sources that minimize destruction or degradation of the environment, though large tradeoffs are always made with today's energy sources.

Even though SPS is technically feasible, environmentally attractive, and electricity is highly flexible and clean, will SPS be economically competitive? People have worked in the construction industry, and also intensely studied different energy supply technologies, and can produce this basic, common sense analysis.

> "Compared to today's energy sources, the SPS and rectenna system offers an economically competitive large scale energy source, and in fact appears to offer a much less expensive energy source once significant space-based infrastructure is established. In addition, the SPS and rectenna system has strong advantages in terms of environmental issues. It appears that SPS will eventually become the premier energy source for Earth."

The SPS is compared to both conventional energy sources (fossil fuels and nuclear), as well as alternative energy sources. The SPS is solar energy, so that it falls under the category of alternative energy. Compared to other solar energy concepts to date, the SPS is clearly the most feasible long-term, large scale solar power source for our economies, as well as the most economical.

SPS Compared to today's Energy Sources

Currently, the world gets about 95% of its energy from coal, oil and natural gas, and almost all of the other 5% from nuclear power and hydroelectric dams.

The SPS concept appears to have inherent promise to be a most economical source of electric power to our economies, relative to today's electricity sources and all other energy sources seriously projected for the forseeable future.

The economics of energy supply consists of two elements:

- The cost of building the power plant and its supporting infrastructure, i.e., the front end capital cost, and
- The cost of operating the power plant, e.g., mines, fuel processing facilities, transportation infrastructure, and waste disposal.

Let's first look at a coal fired power plant for comparison. In terms of front end capital costs, a coal fired power plant has massive generators and mechanical systems constructed from a great variety of precision parts which are expensive. A SPS consists of a very small variety of simple parts mass produced in great quantity.

A SPS needs no fuel supply and waste handling systems.

The SPS consists mainly of a flat plane of silicon solar cells on a beam structure, with electric busbars feeding a large antenna with waveguides and tubes. The satellite is made mostly from a small variety of simple parts mass produced in space. The satellite has only small quantities of precision or special parts.

The receiving antenna on Earth would be mainly screens, posts, and concrete. Practically no moving parts.

The mass of the rectenna would be a little bit less that a coal fired power plant with the same output, assuming a safe, low power density SPS beam, and the satellite in space is about a tenth the mass of a coal fired power plant, to give you a picture of what we're dealing with.

Once built, the SPS and rectenna would continuously supply energy passively with no pollution. In contrast, a coal fired plant of equal power output to an SPS would have to burn tonnages of coal in excess of 20 TIMES the combined weight of the SPS and its ground-based rectenna, and also mine, transport, process and dispose of the ash of these tonnages, each and every year!! This is a massively expensive operation, yet it is the least expensive electricity source today which can reliably supply electrical energy in quantities large enough for our demands.

A nuclear power plant is much more complex than a coal plant, i.e., composed of an even greater variety of specialized and expensive components. A nuclear power plant is about twice as massive as a rectenna, considering just the power plant and not all the facilities required for nuclear fuel mining, transport, purification, enrichment, rod fabrication, spent fuel temporary storage, processing and disposal facilities. Nuclear power plants have very high front-end capital costs (especially with ever-changing safety regulations and the need for nuclear safety), but lower operating costs compared to coal-fired plants. Nuclear power also has long pending nuclear waste disposal issues.

Source	Cost per megawatt	Suppliable on large scale?	Environmentally challenging?	Versatility	Limits
Coal	low	yes	acid rain, CO_2, mining, waste	electric	~none
Oil	low (pre-peak)	yes (peaks ~2010)	acid rain, CO_2, spills	transport	exhaustible
Natural gas	low (pre-peak)	yes (peaks ~2025)	CO_2, some acid raid	electric, heat, LNG for trans-port	~exhaustible
Hydroelectric	low	no (~4%)	flooding	electric	econ. sites
Nuclear fission	low	yes	radiation, terrorism	electric	none
Fusion	high	far future	radiation	electric	none
Photovoltaic (ground-based)	medium	sunny day-times (unreliable)	OK	electric	sunny, south
Solar space heating	low	n/a	OK	space heat	sunny winters, new structures
Solar thermal electric (ground)	medium	sunny day-times (unreliable)	OK	electric	sunny south
Solar thermal ind. heat (ground)	medium	sunny day-times (unreliable)	OK	thermal	sunny south
Wind energy	OK	no	loud	electric	some coasts
Alcohol fuels	livable	OK	OK	transport	~OK
Biomass gas	livable	no	OK	gas	~OK
Ocean thermal	high	no	OK	electric	OK
Geothermal	medium	no	brine, sulfur, toxic metals	electric	few sites

The most important case against adding more nuclear fission power plants is that it takes only 20 kilograms (45 pounds) of plutonium, about the size of a grapefruit, to produce a bomb. It's been 50 years since the first nuclear bomb was made, and advanced technical knowhow is now widespread. Much smaller amounts of plutonium are sufficient for a terrorist to fatally poison areas. The more nuclear plants there are, the greater the chances of an offensive person obtaining the material one way or another in the fuel cycle. Nuclear fission is a risky energy source for the world's growing energy needs, and the sooner we start moving away from it, the better.

Coal and nuclear are more economical and reliable large scale power sources than alternative energy sources, which is the bottom line of why they are prevalent. The above table below summarizes the weak points of alternative energy, e.g., cost per unit of power produced, whether there's enough of it to power a significant percentage of our electricity needs, environmental problems, versatility (e.g., passive solar space heating reduces energy consumption but does not provide other energy so it's a partial solution), and location of the energy source relative to demand.

Nuclear fusion using hydrogen will be technically feasible someday, but it does not look any more economical at this point in time than fission or coal. The power plants will be complex. Plus, fusion is not clean. Fusion produces dangerous radioactive gases, e.g., tritium hydrogen, which are much harder to contain than the solid waste products of nuclear fission. Hydrogen is the most difficult element of all to contain. These radioactive gases must be contained during production (breeding) and extraction, purification, plasma fueling and of course disposal of radioactive wastes and ancillary equipment.

If we spent the same amount of money on SPSs as we do on oversold fusion, we would have a more economical and environmentally clean alternative energy source, sooner. There are no significant advances in technology required for SPSs, in contrast to the advances in technology required for fusion. The spinoffs of space development far outweigh the spinoffs due to fusion power.

You can see that the SPS and rectenna have some natural economic advantages over conventional energy sources, once space based infrastructure in emplaced. Convincing arguments can be made that the SPS will be the main power source for Earth from an economic standpoint in the long-term future.

Compared to Alternative Power Sources

Unlike other solar energy concepts, the SPS would supply solar energy 24 hours per day. It would be reliable, too. Unlike the SPS, ground-based solar energy can't supply power at night unless it has expensive storage equipment plus extra generating capacity during the day. Further, power from terrestrial solar power plants requires backup power for cloudy days, and power generated and stored varies with the seasons.

Unreliability and planning needs make conventional solar energy inconvenient and unattractive to responsible utility companies - an economy can't be held hostage to the whims of the weather.

Reasons why "alternative energy" concepts like wind power and ground-based solar cells have not caught on is because each concept suffers from one or more of the following issues:

- Abundance.
- Concentration of energy supply.
- Reliability.
- Cost per kilowatt.

Regarding reliability, an economy can't shut down because the wind stops blowing the wind power generators as much, causing demand/supply fluctuations in energy prices, rationing or power failures. Also, it takes a lot of wind generators in a windy place to power a small city, for example.

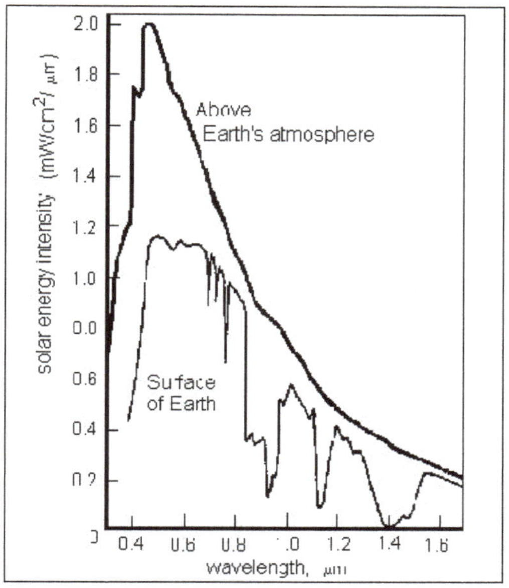

The effects of atmospheric absorption of solar energy on Earth on a fairly clear day. Of course, SPSs in space do not suffer from any atmospheric absorption.

The figure below shows the seasonal variations of solar energy on the surface of the Earth. Again, the SPS is not affected.

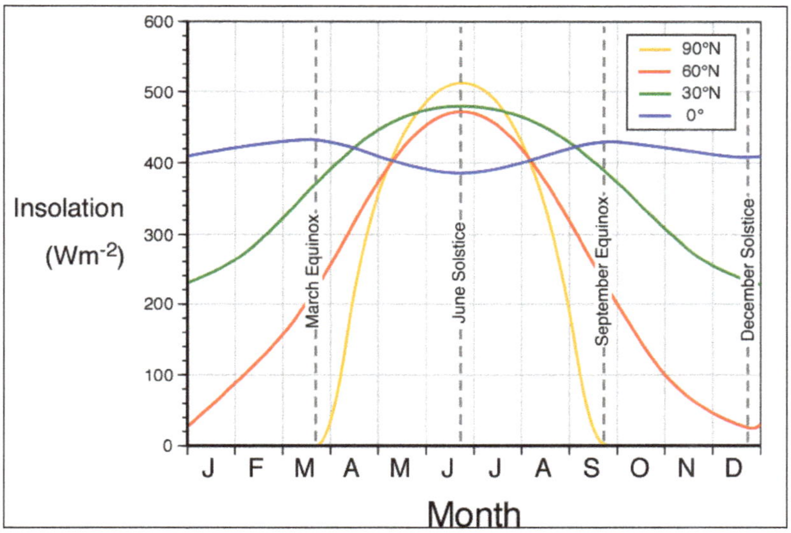

Geothermal energy, hydroelectric dams, and other renewable sources of energy exist, but there is not nearly enough of it to power our economies, especially at costs nearly competitive to fossil fuels. In contrast, the SPS is an abundant, reliable and a natural 24-hour supplier of energy, and rectennas can be located within sufficient proximity to consumers everywhere.

Oil, Gas and the Electric Economy Concepts

The SPS poses a clear alternative to coal and nuclear power plants. But what about oil and natural gas? Can the SPS reduce the vulnerability of the world economy to oil cutoffs (e.g., due to a Middle

East war, terrorism, or embargoes)? Would it be wise to divert our budgets away from wasteful military hardware and invest it into space development (by direct government subsidization and/or massive tax incentives)? Do we need a lead time well before oil supply starts to fall short of oil demand?

The "Electric Economy" concepts - electric heat, electric vehicles, synthetic liquid fuels made with the help of electrical energy - would reduce the growing world economies' vulnerability to energy shortages. This means using clean electrical energy in place of natural gas heating, and making synthetic liquid fuels from natural gas, coal, and hydrogen gas from water electrolysis.

Electric cars would be more popular if parking lots at work, shopping centers, etc., were equipped with simple plug-in recharge meters. Electric cars are clean and quiet.

Liquid fuel for long range vehicles is the main energy source which SPS electricity cannot substitute for directly. However, it can substitute indirectly by providing energy in making synthetic fuels. Also, by using electrical energy in place of oil and natural gas wherever possible, oil and natural gas are liberated for use in long range vehicles.

This requires a willingness to change a little of our infrastructure for the better, e.g., setting up plugs for electric vehicle chargers in parking lots (e.g., office buildings, shopping malls, restaurants, etc.), and opening quick-battery-swapping service stations on the road. Pumping petrol will become a thing of the past for those who choose electric vehicles.

Analyses of the costs of electric vehicles reveals that the cost of electric cars would be roughly the same as present day automobiles in a scenario of mass production of electric vehicles anywhere near the current production of gasoline (petrol) cars. The electric cars are currently more expensive because of the cost of the batteries (front-end capital cost), but the operating costs of electric cars is less since electricity costs less than gasoline, and less maintenance is required on an electric vehicle than a car powered by an internal combustion engine. Again, a main impediment is that there's no plug to charge your car at the office parking lot, at the shopping center, at restaurants, etc.

It's worth noting that from the 1920s through World War II, oil-starved Germany made synthetic gasoline on a massive scale from coal and natural gas, enough to supply the massive Nazi military machine almost exclusively.

Synthetic liquid fuels are made by hydrogenating coal and/or getting natural gas to bond into bigger molecules to form liquids. Petrol (i.e., gasoline) consists of hydrocarbons, i.e., molecules made up of carbon chains surrounded by hydrogen. Coal is pure carbon. To produce synthetic fuels from coal, you combine it with hydrogen. The hydrogen can be supplied by using electricity to split up water into hydrogen and oxygen. In the late 1970s, the U.S. funded synthetic fuels research in response to the Arab oil cutoffs, but those budget items were cut in the early 1980s, which coincided with the time an international oil shortage became an oil glut. Today, we rely on military spending and realpolitik to protect the world economies from oil cutoffs.

Oil makes up 40% of modernised countries' energy consumption. Western Europe, Japan, and the U.S. (WEJUS) consume more than 60 million barrels each and every day! We can readily replace city car gasoline consumption by using electric cars instead. By doing so, we need replace only 12 of the 60 million barrels per day equivalent in electricity, because electric cars are more than 80% efficient whereas gasoline cars are only 15% efficient (yes, most of the energy goes out the tailpipe as heat instead of motion).

Electric cars of range approximately 100 miles (150 km) at highway power without recharge are just recently starting to becomeeconomically attractive to consumers. Charging at parking lots can readily be done, though this kind of infrastructure is hard to get started, and is a great barrier to realization - a chicken-or-egg situation.

However, we are eventually going to have to face the fact that fossil fuel production can't keep up with growing demand for fossil fuel energy because of both depletion of resources and the increased demand due to rapid development of less developed countries. Tax incentives and other measures may need to be emplaced by governments to deal with the situation, to the detriment of our future economic fluidity.

It's a simple fact that new oil discoveries have not kept pace with consumption. In the relatively oil-rich USA, oil reserves have fallen steadily since the 1960s despite technological improvements in exploration and drilling plus significantly increased drilling rates. The US imports more than 50% of its oil, today more than at any other time in history, and US dependence on imported oil only continues to increase. Europe imports more than 70%, and Japan and most free Asian nations import nearly 100% of their oil.

Modern analyses project that profitable synthetic gasoline would go to market at about $2 per gallon using slightly improved German methods from the 1920s - about twice the price of today's gasoline in the US, or about the current price in Japan and Europe where gasoline is heavily taxed to discourage overconsumption. Synthetic fuel is not produced anywhere since Middle East oil is so much cheaper and drives synthetic fuels out of the market at the present time.

There is no place on Earth like the Persian Gulf in terms of oil abundance and cheap production.

As the rest of the world grows, we can expect the Free World to become increasingly dependent on the volatile Middle East for its oil, and thus potentially hostage to a cutoff for any reason. If and when that happens, you can expect worldwide economic recession and hardships, and excuses for outside armies to go in and try to get the oil flowing back out (and which may not be successful despite strong efforts and presences), resulting in the vast spread of terrorism, perhaps with plutonium in hand.

We can expect to maintain high military spending. However, taking a small fraction of military spending and putting it into SPS development would ultimately enhance our security permanently, and be much more economically productive and positive.

Even a beefed up military will not be enough one day, as growing world demand for oil will eventually outstrip supply.

We've been tapping most of the best oil producing geologies for decades, and more than 100 million barrels of oil each and every day is a lot of pumping from the Earth. Increasing demand from newly industrializing countries will eventually cause stiff competition for oil and price rises.

However, production of synthetic liquid fuels from coal is polluting and poses formidable environmental challenges.

We are going to have to find an alternative to consuming massive quantities of oil. If we cut out a big chunk of oil consumption by replacing half our city cars with electric cars, we will significantly relieve consumption of fossil fuels and synthetic fuels.

Putting off the problem until later only makes the problem worse. The sooner we make a committment to solving this problem, the slower we will consume fossil fuels, the less vulnerable our interdependent economies will become, and the better our futures will be. Better sooner than later.

Conservation can help, but it's not the answer. The bulk of the growth in oil consumption is coming from the less developed countries which are industrializing. It's bad enough to see the US, Japan and Western Europe alone consuming around 80 million barrels or more of oil every day for the indefinite future. When you see the industrialization of the other 80% of the world, you get an idea of the magnitude of the problem.

The applications of SPSs are immediately apparent in less developed countries like this. Instead of building deep sea ports for ocean tankers, and so many refineries, pipelines, and petrol stations, for handling imports of Middle East oil, they could be supporting SPS development with the energy piped in by simple electric power lines. They would pay less for electricity than for oil, thus improving their trade balance. Electrical power lines are less destructive to the environment than the alternatives.

Instead of trying to increase domestic oil and gas production by offshore drilling, which is polluting the beautiful beaches and unique underwater sealife, long-term planning could consider substitutions of electrical energy. Already, countries like Thailand have suffered great environmental damage downwind from coal fired power plants and downstream from coal mines, and is seriously considering nuclear power beyond Thailand's current small research generators. Electricity is expensive here. Unlike in industrialized countries, liquified natural gas is popular here, largely from offshore drilling, and used for cooking and many new taxis (as in Australia).

Again, better to offer SPSs sooner than later.

In poor countries where people struggle to survive, environmental preservation is a luxury. Alternatives to environmentally destructive lifestyles must be economically available, feasible and attractive.

Electricity is a most convenient, flexible and clean modern energy source.

The rectennas are simple, relatively cheap construction items, requiring little technical know-how, and can have dual use as irrigation structures or fish farming upon bodies of water. Development around SPS electricity would be much more financially attractive for developing countries to importing oil and coal, compared to the alternatives.

It seems that positive natural market mechanisms will come to bear worldwide when we make progress on SPS development, as well as political decisions. Really, if Thailand is considering nuclear power, don't you think they would prefer solar power satellites and rectennas?

As for non-market mechanisms, it's worth noting that the US spends 300 billion dollars a year on its military to police the world, while practically nothing is spent on developing energy alternatives.

If instead, tax incentives were budgeted for private sector space development of SPS, with perhaps government subsidation (e.g., matching funds for leading companies or consortiums for the first few years just to speed up the development process), we would all get a whole lot more for our money. It's win-win for everyone in every country, especially compared to the alternatives. It will also be the advent of new energy exporters - of solar power, up in space far from political radicals.

In any case, it seems fairly clear that SPSs and rectennas will be a major long-term power source of choice all over the world, and whatever entity develops the source and gets the patents will create a wealthy, glorious and historic legacy.

SOLAR PANELS ON SPACECRAFT

Spacecraft operating in the inner Solar System usually rely on the use of photovoltaic solar panels to derive electricity from sunlight. In the outer solar system, where the sunlight is too weak to produce sufficient power, radioisotope thermoelectric generators (RTGs) are used as a power source.

A solar panel array of the International Space Station.

Uses

The solar panels on the SMM satellite provided electrical power. Here it is being captured by an astronaut in a mobile space-suit that runs on chemical battery power.

Solar panels on spacecraft supply power for two main uses:

- Power to run the sensors, active heating, cooling and telemetry.
- Power for spacecraft propulsion – electric propulsion, sometimes called solar-electric propulsion.

For both uses, a key figure of merit of the solar panels is the specific power (watts generated divided by solar array mass), which indicates on a relative basis how much power one array will generate for a given launch mass relative to another. Another key metric is stowed packing efficiency (deployed watts produced divided by stowed volume), which indicates how easily the array will fit into a launch vehicle. Yet another key metric is cost (dollars per watt).

To increase the specific power, typical solar panels on spacecraft use close-packed solar cell rectangles that cover nearly 100% of the sun-visible area of the solar panels, rather than the solar wafer circles which, even though close-packed, cover about 90% of the sun-visible area of typical solar panels on earth. However, some solar panels on spacecraft have solar cells that cover only 30% of the sun-visible area.

Implementation

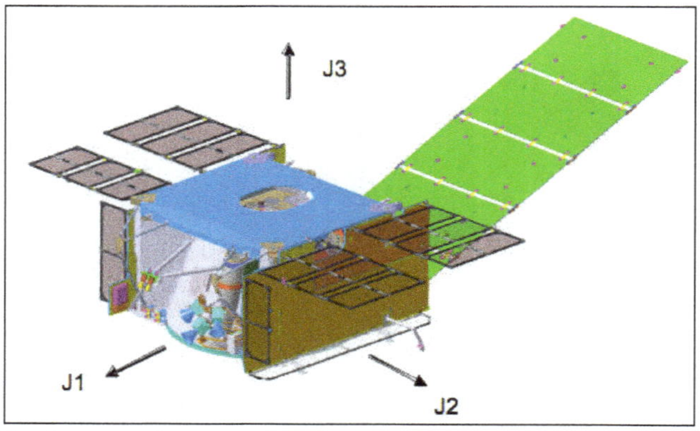

Diagram of the Spacecraft Bus on the planned James Webb Space Telescope, which is powered by solar panels (coloured green in this 3/4 view). Note that shorter light purple extensions are radiator shades not solar panels.

Solar panels need to have a lot of surface area that can be pointed towards the Sun as the spacecraft moves. More exposed surface area means more electricity can be converted from light energy from the Sun. Since spacecraft have to be small, this limits the amount of power that can be produced.

All electrical circuits generate waste heat; in addition, solar arrays act as optical and thermal as well as electrical collectors. Heat must be radiated from their surfaces. High-power spacecraft may have solar arrays that compete with the active payload itself for thermal dissipation. The innermost panel of arrays may be "blank" to reduce the overlap of views to space. Such spacecraft include the higher-power communications satellites (e.g., later-generation TDRS) and Venus Express, not high-powered but closer to the Sun.

Spacecraft are built so that the solar panels can be pivoted as the spacecraft moves. Thus, they can always stay in the direct path of the light rays no matter how the spacecraft is pointed. Spacecraft are usually designed with solar panels that can always be pointed at the Sun, even as the rest of the body of the spacecraft moves around, much as a tank turret can be aimed independently of where the tank is going. A tracking mechanism is often incorporated into the solar arrays to keep the array pointed towards the sun.

Sometimes, satellite operators purposefully orient the solar panels to "off point," or out of direct alignment from the Sun. This happens if the batteries are completely charged and the amount of electricity needed is lower than the amount of electricity made; off-pointing is also sometimes used on the International Space Station for orbital drag reduction.

Ionizing Radiation Issues and Mitigation

Juno is the second spacecraft to orbit Jupiter and the first solar-powered craft to do so.

Space contains varying levels of ionizing radiation, that which includes flares and other solar events. Some satellites orbit within the protective zone of the magnetosphere, while others do not.

Types of Solar Cells Typically used

Gallium arsenide-based solar cells are typically favored over crystalline silicon in industry because they have a higher efficiency and degrade more slowly than silicon in the radiation present in space. The most efficient solar cells currently in production are multi-junction photovoltaic cells. These use a combination of several layers of gallium arsenide, indium gallium phosphide, and germanium to capture more energy from the solar spectrum. Leading edge multi-junction cells are capable of exceeding 38.8% under non-concentrated AM1.5G illumination and 46% using concentrated AM1.5G illumination.

Spacecraft that have used Solar Power

To date, solar power, other than for propulsion, has been practical for spacecraft operating no farther from the Sun than the orbit of Jupiter. For example, Juno, Magellan, Mars Global Surveyor, and Mars Observer used solar power as does the Earth-orbiting, Hubble Space Telescope. The Rosetta space probe, launched 2 March 2004, used its 64 square metres (690 sq ft) of solar panels as far as the orbit of Jupiter (5.25 AU); previously the furthest use was the Stardust spacecraft at

2 AU. Solar power for propulsion was also used on the European lunar mission SMART-1 with a Hall effect thruster.

The Juno mission, launched in 2011, is the first mission to Jupiter (arrived at Jupiter on July 4, 2016) to use solar panels instead of the traditional RTGs that are used by previous outer solar system missions, making it the furthest spacecraft to use solar panels to date. It has 72 square metres (780 sq ft) of panels.

Another spacecraft of interest is Dawn which went into orbit around 4 Vesta in 2011. It used ion thrusters to get to Ceres.

The potential for solar powered spacecraft beyond Jupiter has been studied.

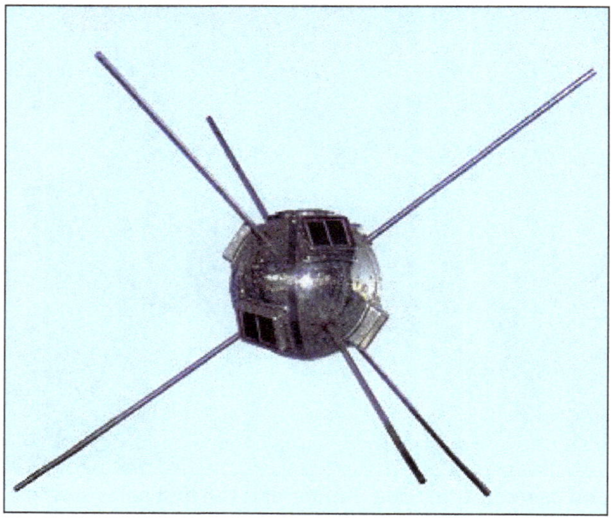

Vanguard 1, the first solar powered satellite.

The International Space Station also uses solar arrays to power everything on the station. The 262,400 solar cells cover around 27,000 square feet (2,500 m^2) of space. There are four sets of solar arrays that power the station and the fourth set of arrays were installed in March 2009. 84 to 120 kilowatts of electricity can be generated from these solar arrays.

Rosetta's lander Philae.

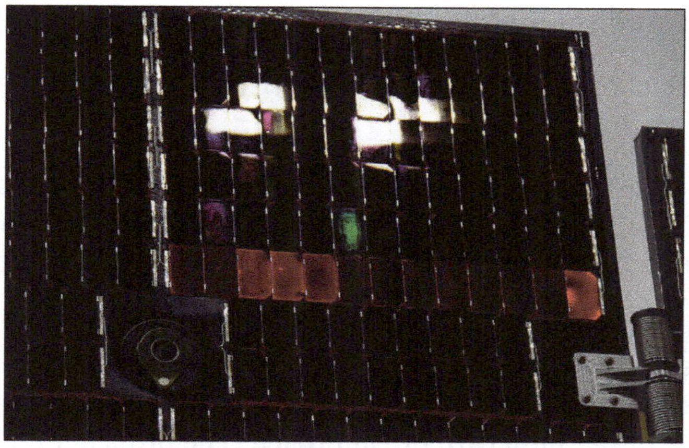

A section of one of Juno's solar panels.

The Juno space probe.

SOLAR THERMAL ROCKET

A solar thermal rocket is a theoretical spacecraft propulsion system that would make use of solar power to directly heat reaction mass, and therefore would not require an electrical generator, like most other forms of solar-powered propulsion do. The rocket would only have to carry the means of capturing solar energy, such as concentrators and mirrors. The heated propellant would be fed through a conventional rocket nozzle to produce thrust. Its engine thrust would be directly related to the surface area of the solar collector and to the local intensity of the solar radiation.

In the shorter term, solar thermal propulsion has been proposed both for longer-life, lower-cost, more efficient use of the sun and more-flexible cryogenic upper stage launch vehicles and for on-orbit propellant depots. Solar thermal propulsion is also a good candidate for use in reusable inter-orbital tugs, as it is a high-efficiency low-thrust system that can be refuelled with relative ease.

Solar-thermal Design Concepts

There are two solar thermal propulsion concepts, differing primarily in the method by which they use solar power to heat up the propellant:

- Indirect solar heating involves pumping the propellant through passages in a heat exchanger that is heated by solar radiation. The windowless heat exchanger cavity concept is a design taking this radiation absorption approach.

- Direct solar heating involves exposing the propellant directly to solar radiation. The rotating bed concept is one of the preferred concepts for direct solar radiation absorption; it offers higher specific impulse than other direct heating designs by using a retained seed (tantalum carbide or hafnium carbide) approach. The propellant flows through the porous walls of a rotating cylinder, picking up heat from the seeds, which are retained on the walls by the rotation. The carbides are stable at high temperatures and have excellent heat transfer properties.

Due to limitations in the temperature that heat exchanger materials can withstand (approximately 2800 K), the indirect absorption designs cannot achieve specific impulses beyond 900 seconds (9 kN·s/kg = 9 km/s) (or up to 1000 seconds). The direct absorption designs allow higher propellant temperatures and therefore higher specific impulses, approaching 1200 seconds. Even the lower specific impulse represents a significant increase over that of conventional chemical rockets, however, an increase that can provide substantial payload gains (45 percent for a LEO-to-GEO mission) at the expense of increased trip time (14 days compared to 10 hours).

Small-scale hardware has been designed and fabricated for the Air Force Rocket Propulsion Laboratory (AFRPL) for ground test evaluation. Systems with 10 to 100 N of thrust have been investigated by SART.

Reusable Orbital Transfer Vehicles (OTV), sometimes called (inter-orbital) space tugs, propelled by solar thermal rockets have been proposed. The concentrators on solar thermal tugs are less susceptible to radiation in the Van Allen belts than the solar arrays of solar electric OTV.

Propellants

Most proposed designs for solar thermal rockets use hydrogen as their propellant due to its low molecular weight which gives excellent specific impulse of up to 1000 seconds (10 kN·s/kg) using heat exchangers made of rhenium.

Conventional thought has been that hydrogen—although it gives excellent specific impulse—is not space storable. Design work in the early 2010s has developed an approach to substantially reduce hydrogen boiloff, and to economically utilize the small remaining boiloff product for requisite in-space tasks, essentially achieving zero boil off (ZBO) from a practical point of view.

Other substances could also be used. Water gives quite poor performance of 190 seconds (1.9 kN·s/kg), but requires only simple equipment to purify and handle, and is space storable and this has very seriously been proposed for interplanetary use, using in-situ resources.

Ammonia has been proposed as a propellant. It offers higher specific impulse than water, but is easily storable, with a freezing point of −77 degrees Celsius and a boiling point of −33.34 °C. The exhaust dissociates into hydrogen and nitrogen, leading to a lower average molecular weight, and thus a higher Isp (65% of hydrogen).

A solar-thermal propulsion architecture outperforms architectures involving electrolysis and liquification of hydrogen from water by more than an order of magnitude, since electrolysis requires heavy power generators, whereas distillation only requires a simple and compact heat source (either nuclear or solar); so the propellant production rate is correspondingly far higher for any given initial mass of equipment. However its use does rely on having clear ideas of the location of water ice in the solar system, particularly on lunar and asteroidal bodies, and such information is not known, other than that the bodies with the asteroid belt and further from the Sun are expected to be rich in water ice.

Solar-thermal for Ground Launch

Solar thermal rockets have been proposed as a system for launching a small personal spacecraft into orbit. The design is based on a high altitude airship which uses its envelope to focus sunlight onto a tube. The propellant, which would likely be ammonia, is then fed through to produce thrust. Possible design flaws include whether the engine could produce enough thrust to overcome drag, and whether the skin of the airship wouldn't fail at hypersonic velocities. This has many similarities to the orbital airship proposed by JP Aerospace.

Proposed Solar-thermal Space Systems

As of 2010, two proposals for utilizing solar-thermal propulsion on in-space post-launch spacecraft systems had been made.

A concept to provide low Earth orbit (LEO) propellant depots that could be used as way-stations for other spacecraft to stop and refuel on the way to beyond-LEO missions has proposed that waste gaseous hydrogen—an inevitable byproduct of long-term liquid hydrogen storage in the radiative heat environment of space—would be usable as a monopropellant in a solar-thermal propulsion system. The waste hydrogen would be productively utilized for both orbital stationkeeping and attitude control, as well as providing limited propellant and thrust to use for orbital maneuvers to better rendezvous with other spacecraft that would be inbound to receive fuel from the depot.

Solar-thermal monoprop hydrogen thrusters are also integral to the design of the next-generation cryogenic upper stage rocket proposed by U.S. company United Launch Alliance (ULA). The Advanced Common Evolved Stage (ACES) was intended as a lower-cost, more-capable and more-flexible upper stage that would supplement, and perhaps replace, the existing ULA Centaur and ULA Delta Cryogenic Second Stage (DCSS) upper stage vehicles. The ACES Integrated Vehicle Fluids option eliminates all hydrazine monopropellant and all helium pressurant from the space vehicle—normally used for attitude control and station keeping—and depends instead on solar-thermal monoprop thrusters using waste hydrogen.

The viability of various trips using Solar Thermal propulsion was investigated by Gordon Woodcock and Dave Byers in 2003.

A subsequent proposal in the 2010s was the Solar Moth spacecraft that would use lightweight mirrors to focus solar radiation on a solar thermal engine.

SOLAR ELECTRIC PROPULSION

Solar electric propulsion (SEP) typically refers to the combination of solar cells and an ion drive to propel a spacecraft through outer space. This technology has been exploited in a variety of spacecraft by the European Space Agency, the Japanese Space Agency, Indian Space Research Organisation and NASA. SEP is valued for its potential for a high specific impulse, thus allowing less mass for propellant to be launched with a spacecraft and it has been evaluated for mission to Mars.

View of Deep Space 1, showing both the solar panels and ion engine (with blue exhaust), major aspects of this solar electric design. Solar energy may also be temporarily stored in chemical battery inside the spacecraft bus.

The Dawn spacecraft's Xenon tank prior to integration with spacecraft. The Xenon was the propellant for the solar-power ion drive of the spacecraft which would go on to orbit two different asteroids in the early 21st century.

Roll-out solar panel tested in Earth Orbit at the International Space Station, 2017.

Solar electric propulsion combines solar panels on spacecraft and ion thrusters, used in tandem. However, there are a number of other types of electrically powered spacecraft propulsion such as arcjet rockets. It may not employ an ion drive, of which there is a variety of types also.

It is also possible to generate electricity from the Sun without using photovoltaic panels, such as with solar concentrators and a Stirling engine.

A 50 kilowatt SEP system was studied in the 2010s for a mission to an asteroid. In February 2012 NASA awarded a contract for a Solar Electric Propulsion Flight System.

An example of work on this type of technology is Advanced Electric Propulsion System.

The NASA Solar Technology Application Readiness (NSTAR) ion engine has been used with photovoltaic solar panels, which was tested on the Deep Space 1 mission along with Solar Concentrator Arrays (Launched in 1998 as part of the New Millennium Program).

SEP has been studied as a technology for a mission to Mars. In particular the high specific impulse of the ion engines could lower overall mass and avoid having to use nuclear technology for power when coupled with solar panels. A 1998 study for SEP for a human mission suggest that a human-sized spacecraft would need 600 to 800 kilowatts of electrical power coupled with ion engines with a specific impulse of 2000 to 2500 seconds.

SOLAR SAIL

Solar sails (also called light sails or photon sails) are a proposed method of spacecraft propulsion using radiation pressure exerted by sunlight on large mirrors. A useful analogy may be a sailing boat; the light exerting a force on the mirrors is akin to a sail being blown by the wind. High-energy laser beams could be used as an alternative light source to exert much greater force than would be possible using sunlight, a concept known as beam sailing.

IKAROS space-probe with solar sail in flight (artist's depiction) showing a typical square sail configuration.

Solar sail craft offer the possibility of low-cost operations combined with long operating lifetimes. Since they have few moving parts and use no propellant, they can potentially be used numerous times for delivery of payloads.

Solar sails use a phenomenon that has a proven, measured effect on spacecraft. Solar pressure affects all spacecraft, whether in interplanetary space or in orbit around a planet or small body. A typical spacecraft going to Mars, for example, will be displaced thousands of kilometers by solar pressure, so the effects must be accounted for in trajectory planning, which has been done since the time of the earliest interplanetary spacecraft of the 1960s. Solar pressure also affects the orientation of a craft, a factor that must be included in spacecraft design.

The total force exerted on an 800 by 800 meter solar sail, for example, is about 5 newtons (1.1 lbf) at Earth's distance from the Sun, making it a low-thrust propulsion system, similar to spacecraft propelled by electric engines, but as it uses no propellant, that force is exerted almost constantly and the collective effect over time is great enough to be considered a potential manner of propelling spacecraft.

Johannes Kepler observed that comet tails point away from the Sun and suggested that the Sun caused the effect. In a letter to Galileo in 1610, he wrote, "Provide ships or sails adapted to the heavenly breezes, and there will be some who will brave even that void." He might have had the comet tail phenomenon in mind when he wrote those words, although his publications on comet tails came several years later.

James Clerk Maxwell, in 1861–1864, published his theory of electromagnetic fields and radiation, which shows that light has momentum and thus can exert pressure on objects. Maxwell's equations provide the theoretical foundation for sailing with light pressure. So by 1864, the physics community and beyond knew sunlight carried momentum that would exert a pressure on objects.

Jules Verne, in From the Earth to the Moon, published in 1865, wrote "there will some day appear velocities far greater than these [of the planets and the projectile], of which light or electricity will probably be the mechanical agent we shall one day travel to the moon, the planets, and the stars." This is possibly the first published recognition that light could move ships through space.

Pyotr Lebedev was first to successfully demonstrate light pressure, which he did in 1899 with a torsional balance; Ernest Nichols and Gordon Hull conducted a similar independent experiment in 1901 using a Nichols radiometer.

Svante Arrhenius predicted in 1908 the possibility of solar radiation pressure distributing life spores across interstellar distances, providing one means to explain the concept of panspermia. He apparently was the first scientist to state that light could move objects between stars.

Konstantin Tsiolkovsky first proposed using the pressure of sunlight to propel spacecraft through space and suggested, "using tremendous mirrors of very thin sheets to utilize the pressure of sunlight to attain cosmic velocities".

Friedrich Zander (Tsander) published a technical paper in 1925 that included technical analysis of solar sailing. Zander wrote of "applying small forces" using "light pressure or transmission of light energy to distances by means of very thin mirrors".

JBS Haldane speculated in 1927 about the invention of tubular spaceships that would take humanity to space and how "wings of metallic foil of a square kilometre or more in area are spread out to catch the Sun's radiation pressure".

J. D. Bernal wrote in 1929, "A form of space sailing might be developed which used the repulsive effect of the Sun's rays instead of wind. A space vessel spreading its large, metallic wings, acres in extent, to the full, might be blown to the limit of Neptune's orbit. Then, to increase its speed, it would tack, close-hauled, down the gravitational field, spreading full sail again as it rushed past the Sun."

The first formal technology and design effort for a solar sail began in 1976 at Jet Propulsion Laboratory for a proposed mission to rendezvous with Halley's Comet.

Physical Principles

Solar Radiation Pressure

Many people believe that spacecraft using solar sails are pushed by the Solar winds just as sailboats and sailing ships are pushed by the winds across the waters on Earth. But Solar radiation exerts a pressure on the sail due to reflection and a small fraction that is absorbed.

The momentum of a photon or an entire flux is given by Einstein's relation:

$$p = E/c$$

where p is the momentum, E is the energy (of the photon or flux), and c is the speed of light. Specifically the momentum of a photon depends on its wavelength $p = h/\lambda$.

Solar radiation pressure can be related to the irradiance (solar constant) value of 1361 W/m^2 at 1 AU (Earth-Sun distance), as revised in 2011:

- Perfect absorbance: F = 4.54 μN per square metre (4.54 μPa) in the direction of the incident beam (an inelastic collision)
- Perfect reflectance: F = 9.08 μN per square metre (9.08 μPa) in the direction normal to surface (an elastic collision)

An ideal sail is flat and has 100% specular reflection. An actual sail will have an overall efficiency of about 90%, about 8.17 $\mu N/m^2$, due to curvature (billow), wrinkles, absorbance, re-radiation from front and back, non-specular effects, and other factors.

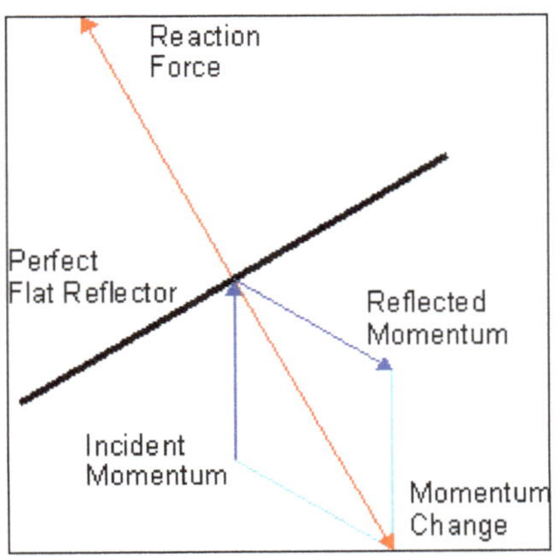

Force on a sail results from reflecting the photon flux.

The force on a sail and the actual acceleration of the craft vary by the inverse square of distance from the Sun (unless extremely close to the Sun), and by the square of the cosine of the angle between the sail force vector and the radial from the Sun. So,

$$F = F_o \cos^2 \theta / R^2 \text{ (ideal sail)}$$

where R is distance from the Sun in AU. An actual square sail can be modeled as:

$$F = F_o (0.349 + 0.662 \cos 2\theta - 0.011 \cos 4\theta)/R^2$$

Note that the force and acceleration approach zero generally around $\theta = 60°$ rather than $90°$ as one might expect with an ideal sail.

If some of the energy is absorbed, the absorbed energy will heat the sail, which re-radiates that energy from the front and rear surfaces, depending on the emissivity of those two surfaces.

Solar wind, the flux of charged particles blown out from the Sun, exerts a nominal dynamic pressure of about 3 to 4 nPa, three orders of magnitude less than solar radiation pressure on a reflective sail.

Sail Parameters

Sail loading (areal density) is an important parameter, which is the total mass divided by the sail area, expressed in g/m². It is represented by the Greek letter σ.

A sail craft has a characteristic acceleration, a_c, which it would experience at 1 AU when facing the Sun. Note this value accounts for both the incident and reflected momentums. Using the value from above of 9.08 μN per square metre of radiation pressure at 1 AU, a_c is related to areal density by:

$$a_c = 9.08 \text{ (efficiency)}/\sigma \text{ mm/s}^2$$

Assuming 90% efficiency, $a_c = 8.17/\sigma$ mm/s²

The lightness number, λ, is the dimensionless ratio of maximum vehicle acceleration divided by the Sun's local gravity. Using the values at 1 AU:

$$\lambda = a_c/5.93$$

The lightness number is also independent of distance from the Sun because both gravity and light pressure fall off as the inverse square of the distance from the Sun. Therefore, this number defines the types of orbit maneuvers that are possible for a given vessel.

The table presents some example values. Payloads are not included. The first two are from the detailed design effort at JPL in the 1970s. The third, the lattice sailer, might represent about the best possible performance level. The dimensions for square and lattice sails are edges. The dimension for heliogyro is blade tip to blade tip.

Type	σ (g/m²)	a_c (mm/s²)	λ	Size (km²)
Square sail	5.27	1.56	0.26	0.820
Heliogyro	6.39	1.29	0.22	15
Lattice sailer	0.07	117	20	0.840

Attitude Control

An active attitude control system (ACS) is essential for a sail craft to achieve and maintain a desired orientation. The required sail orientation changes slowly (often less than 1 degree per day) in interplanetary space, but much more rapidly in a planetary orbit. The ACS must be capable of meeting these orientation requirements. Attitude control is achieved by a relative shift between the craft's center of pressure and its center of mass. This can be achieved with control vanes, movement of individual sails, movement of a control mass, or altering reflectivity.

Holding a constant attitude requires that the ACS maintain a net torque of zero on the craft. The total force and torque on a sail, or set of sails, is not constant along a trajectory. The force changes with solar distance and sail angle, which changes the billow in the sail and deflects some elements of the supporting structure, resulting in changes in the sail force and torque.

Sail temperature also changes with solar distance and sail angle, which changes sail dimensions. The radiant heat from the sail changes the temperature of the supporting structure. Both factors affect total force and torque.

To hold the desired attitude the ACS must compensate for all of these changes.

Constraints

In Earth orbit, solar pressure and drag pressure are typically equal at an altitude of about 800 km, which means that a sail craft would have to operate above that altitude. Sail craft must operate in orbits where their turn rates are compatible with the orbits, which is generally a concern only for spinning disk configurations.

Sail operating temperatures are a function of solar distance, sail angle, reflectivity, and front and back emissivities. A sail can be used only where its temperature is kept within its material limits. Generally, a sail can be used rather close to the Sun, around 0.25 AU, or even closer if carefully designed for those conditions.

Applications

Potential applications for sail craft range throughout the Solar System, from near the Sun to the comet clouds beyond Neptune. The craft can make outbound voyages to deliver loads or to take up station keeping at the destination. They can be used to haul cargo and possibly also used for human travel.

Inner Planets

For trips within the inner Solar System, they can deliver loads and then return to Earth for subsequent voyages, operating as an interplanetary shuttle. For Mars in particular, the craft could provide economical means of routinely supplying operations on the planet according to Jerome Wright, "The cost of launching the necessary conventional propellants from Earth are enormous for manned missions. Use of sailing ships could potentially save more than $10 billion in mission costs."

Solar sail craft can approach the Sun to deliver observation payloads or to take up station keeping orbits. They can operate at 0.25 AU or closer. They can reach high orbital inclinations, including polar.

Solar sails can travel to and from all of the inner planets. Trips to Mercury and Venus are for rendezvous and orbit entry for the payload. Trips to Mars could be either for rendezvous or swing-by with release of the payload for aerodynamic braking.

Sail size m	Mercury Rendezvous		Venus Rendezvous		Mars Rendezvous		Mars Aerobrake	
	days	tons	days	tons	days	tons	days	tons
800 $\sigma = 5\ g/m^2$ w/o cargo	600	9	200	1	400	2	131	2
	900	19	270	5	500	5	200	5
	1200	28			700	9	338	10
2000 $\sigma = 3\ g/m^2$ w/o cargo	600	66	200	17	400	23	131	20
	900	124	270	36	500	40	200	40
	1200	184			700	66	338	70

Outer Planets

Minimum transfer times to the outer planets benefit from using an indirect transfer (solar swing-by). However, this method results in high arrival speeds. Slower transfers have lower arrival speeds.

The minimum transfer time to Jupiter for a_c of 1 mm/s^2 with no departure velocity relative to Earth is 2 years when using an indirect transfer (solar swing-by). The arrival speed (V_∞) is close to 17 km/s. For Saturn, the minimum trip time is 3.3 years, with an arrival speed of nearly 19 km/s.

Table: Minimum times to the outer planets (a_c = 1 mm/s^2).

	Jupiter	Saturn	Uranus	Neptune
Time, yr	2.0	3.3	5.8	8.5
Speed, km/s	17	19	20	20

Oort Cloud/Sun's Inner Gravity Focus

The Sun's inner gravitational focus point lies at minimum distance of 550 AU from the Sun, and is the point to which light from distant objects is focused by gravity as a result of it passing by the Sun. This is thus the distant point to which solar gravity will cause the region of deep space on the other side of the Sun to be focused, thus serving effectively as a very large telescope objective lens.

It has been proposed that an inflated sail, made of beryllium, that starts at 0.05 AU from the Sun would gain an initial acceleration of 36.4 m/s^2, and reach a speed of 0.00264c (about 950 km/s) in less than a day. Such proximity to the Sun could prove to be impractical in the near term due to the structural degradation of beryllium at high temperatures, diffusion of hydrogen at high temperatures as well as an electrostatic gradient, generated by the ionization of beryllium from the solar wind, posing a burst risk. A revised perihelion of 0.1 AU would reduce the aforementioned temperature and solar flux exposure. Such a sail would take "Two and a half years to reach the heliopause, six and a half years to reach the Sun's inner gravitational focus, with arrival at the inner Oort Cloud in no more than thirty years." "Such a mission could perform useful astrophysical observations en route, explore gravitational focusing techniques, and image Oort Cloud objects while exploring particles and fields in that region that are of galactic rather than solar origin."

Satellites

Robert L. Forward has commented that a solar sail could be used to modify the orbit of a satellite about the Earth. In the limit, a sail could be used to "hover" a satellite above one pole of the Earth. Spacecraft fitted with solar sails could also be placed in close orbits such that they are stationary with respect to either the Sun or the Earth, a type of satellite named by Forward a "statite". This is possible because the propulsion provided by the sail offsets the gravitational attraction of the Sun. Such an orbit could be useful for studying the properties of the Sun for long durations. Likewise a solar sail-equipped spacecraft could also remain on station nearly above the polar solar terminator of a planet such as the Earth by tilting the sail at the appropriate angle needed to counteract the planet's gravity.

In his book The Case for Mars, Robert Zubrin points out that the reflected sunlight from a large statite, placed near the polar terminator of the planet Mars, could be focused on one of the Martian polar ice caps to significantly warm the planet's atmosphere. Such a statite could be made from asteroid material.

Trajectory Corrections

The MESSENGER probe orbiting Mercury used light pressure on its solar panels to perform fine trajectory corrections on the way to Mercury. By changing the angle of the solar panels relative to the Sun, the amount of solar radiation pressure was varied to adjust the spacecraft trajectory more delicately than possible with thrusters. Minor errors are greatly amplified by gravity assist maneuvers, so using radiation pressure to make very small corrections saved large amounts of propellant.

Interstellar Flight

In the 1970s, Robert Forward proposed two beam-powered propulsion schemes using either lasers or masers to push giant sails to a significant fraction of the speed of light.

In the science fiction novel Rocheworld, Forward described a light sail propelled by super lasers. As the starship neared its destination, the outer portion of the sail would detach. The outer sail would then refocus and reflect the lasers back onto a smaller, inner sail. This would provide braking thrust to stop the ship in the destination star system.

Both methods pose monumental engineering challenges. The lasers would have to operate for years continuously at gigawatt strength. Forward's solution to this requires enormous solar panel arrays to be built at or near the planet Mercury. A planet-sized mirror or fresnel lens would need to be located at several dozen astronomical units from the Sun to keep the lasers focused on the sail. The giant braking sail would have to act as a precision mirror to focus the braking beam onto the inner "deceleration" sail.

A potentially easier approach would be to use a maser to drive a "solar sail" composed of a mesh of wires with the same spacing as the wavelength of the microwaves directed at the sail, since the manipulation of microwave radiation is somewhat easier than the manipulation of visible light. The hypothetical “Starwisp” interstellar probe design would use microwaves, rather than visible light, to push it. Masers spread out more rapidly than optical lasers owing to their longer wavelength, and so would not have as great an effective range.

Masers could also be used to power a painted solar sail, a conventional sail coated with a layer of chemicals designed to evaporate when struck by microwave radiation. The momentum generated by this evaporation could significantly increase the thrust generated by solar sails, as a form of lightweight ablative laser propulsion.

To further focus the energy on a distant solar sail, Forward proposed a lens designed as a large zone plate. This would be placed at a location between the laser or maser and the spacecraft.

Another more physically realistic approach would be to use the light from the Sun to accelerate. The ship would first drop into an orbit making a close pass to the Sun, to maximize the solar energy input on the sail, then it would begin to accelerate away from the system using the light from the Sun. Acceleration will drop approximately as the inverse square of the distance from the Sun, and beyond some distance, the ship would no longer receive enough light to accelerate it significantly, but would maintain the final velocity attained. When nearing the target star, the ship could turn its sails toward it and begin to use the outward pressure of the destination star to decelerate. Rockets could augment the solar thrust.

Similar solar sailing launch and capture were suggested for directed panspermia to expand life in other solar system. Velocities of 0.05% the speed of light could be obtained by solar sails carrying 10 kg payloads, using thin solar sail vehicles with effective areal densities of 0.1 g/m^2 with thin sails of 0.1 μm thickness and sizes on the order of one square kilometer. Alternatively, swarms of 1 mm capsules could be launched on solar sails with radii of 42 cm, each carrying 10,000 capsules of a hundred million extremophile microorganisms to seed life in diverse target environments.

Deorbiting Artificial Satellites

Small solar sails have been proposed to accelerate the deorbiting of small artificial satellites from Earth orbits. Satellites in low Earth orbit can use a combination of solar pressure on the sail and increased atmospheric drag to accelerate satellite reentry. A de-orbit sail developed at Cranfield University is part of the UK satellite TechDemoSat-1, launched in 2014, and is expected to be deployed at the end of the satellite’s five-year useful life. The sail’s purpose is to bring the satellite out of orbit over a period of about 25 years. In July 2015 British 3U CubeSat called DeorbitSail was launched into space with the purpose of testing 16 m^2 deorbit structure, but eventually it failed to deploy it. There is also a student 2U CubeSat mission called PW-Sat2 planned to launch in 2017 that will test 4 m^2 deorbit sail. In June 2017 a second British 3U CubeSat called InflateSail deployed a 10 m^2 deorbit sail at an altitude of 500 kilometers (310 mi). In June 2017 the 3U Cubesat URSAMAIOR has been launched in Low Earth Orbit to test the deorbiting system ARTICA developed by Spacemind. The device, which occupies only 0.4 U of the cubesat, shall deploy a sail of 2.1 m^2 to deorbit the satellite at the end of the operational life.

Sail Configurations

IKAROS, launched in 2010, was the first practical solar sail vehicle. As of 2015, it was still under thrust, proving the practicality of a solar sail for long-duration missions. It is spin-deployed, with tip-masses in the corners of its square sail. The sail is made of thin polyimide film, coated with evaporated aluminium. It steers with electrically-controlled liquid crystal panels. The sail slowly spins, and these panels turn on and off to control the attitude of the vehicle. When on, they diffuse light, reducing the momentum transfer to that part of the sail. When off, the sail reflects more light, transferring more momentum. In that way, they turn the sail. Thin-film solar cells are also integrated into the sail, powering the spacecraft. The design is very reliable, because spin deployment, which is preferable for large sails, simplified the mechanisms to unfold the sail and the LCD panels have no moving parts.

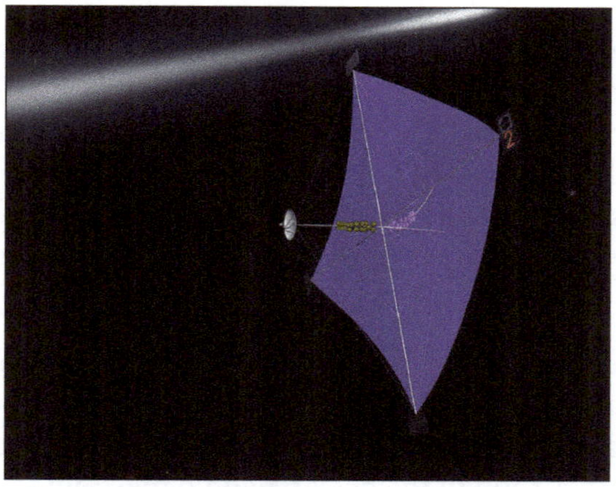

NASA illustration of the unlit side of a half-kilometre solar sail, showing the struts stretching the sail.

An artist's depiction of a Cosmos 1-type spaceship in orbit.

Parachutes have very low mass, but a parachute is not a workable configuration for a solar sail. Analysis shows that a parachute configuration would collapse from the forces exerted by shroud lines, since radiation pressure does not behave like aerodynamic pressure, and would not act to keep the parachute open.

The highest thrust-to-mass designs for ground-assembled deploy-able structures are square sails with the masts and guy lines on the dark side of the sail. Usually there are four masts that spread the corners of the sail, and a mast in the center to hold guy-wires. One of the largest advantages is that there are no hot spots in the rigging from wrinkling or bagging, and the sail protects the structure from the Sun. This form can, therefore, go close to the Sun for maximum thrust. Most designs steer with small moving sails on the ends of the spars.

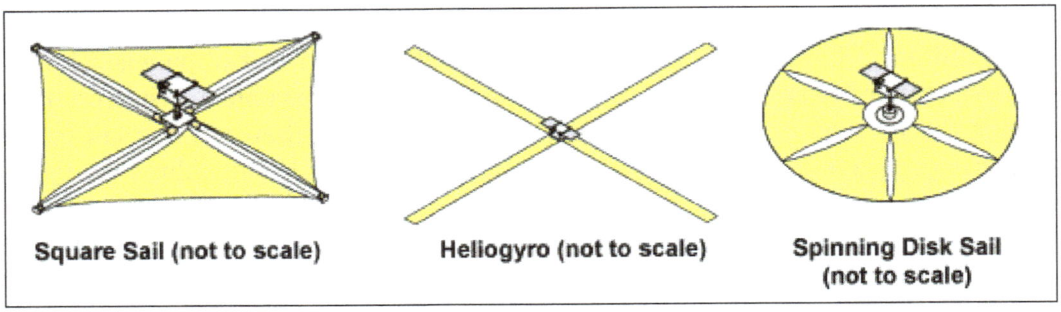

In the 1970s JPL studied many rotating blade and ring sails for a mission to rendezvous with Halley's Comet. The intention was to stiffen the structures using angular momentum, eliminating the need for struts, and saving mass. In all cases, surprisingly large amounts of tensile strength were needed to cope with dynamic loads. Weaker sails would ripple or oscillate when the sail's attitude changed, and the oscillations would add and cause structural failure. The difference in the thrust-to-mass ratio between practical designs was almost nil, and the static designs were easier to control.

JPL's reference design was called the "heliogyro". It had plastic-film blades deployed from rollers and held out by centrifugal forces as it rotated. The spacecraft's attitude and direction were to be completely controlled by changing the angle of the blades in various ways, similar to the cyclic and collective pitch of a helicopter. Although the design had no mass advantage over a square sail, it remained attractive because the method of deploying the sail was simpler than a strut-based design.

Heliogyro design is similar to the blades on a helicopter. The design is faster to manufacture due to lightweight centrifugal stiffening of sails. Also, they are highly efficient in cost and velocity because the blades are lightweight and long. Unlike the square and spinning disk designs, heliogyro is easier to deploy because the blades are compacted on a reel. The blades roll out when they are deploying after the ejection from the spacecraft. As the heliogyro travels through space the system spins around because of the centrifugal acceleration. Finally, payloads for the space flights are placed in the center of gravity to even out the distribution of weight to ensure stable flight.

JPL also investigated "ring sails" (Spinning Disk Sail in the above diagram), panels attached to the edge of a rotating spacecraft. The panels would have slight gaps, about one to five percent of the total area. Lines would connect the edge of one sail to the other. Masses in the middles of these lines would pull the sails taut against the coning caused by the radiation pressure. JPL researchers said that this might be an attractive sail design for large manned structures. The inner ring, in particular, might be made to have artificial gravity roughly equal to the gravity on the surface of Mars.

A solar sail can serve a dual function as a high-gain antenna. Designs differ, but most modify the metalization pattern to create a holographic monochromatic lens or mirror in the radio frequencies of interest, including visible light.

Electric Solar Wind Sail

Pekka Janhunen from FMI has invented a type of solar sail called the electric solar wind sail. Mechanically it has little in common with the traditional solar sail design. The sails are replaced with straightened conducting tethers (wires) placed radially around the host ship. The wires are electrically charged to create an electric field around the wires. The electric field extends a few tens of metres into the plasma of the surrounding solar wind. The solar electrons are reflected by the electric field (like the photons on a traditional solar sail). The radius of the sail is from the electric field rather than the actual wire itself, making the sail lighter. The craft can also be steered by regulating the electric charge of the wires. A practical electric sail would have 50–100 straightened wires with a length of about 20 km each.

Electric solar wind sails can adjust their electrostatic fields and sail attitudes.

Magnetic Sail

A magnetic sail would also employ the solar wind. However, the magnetic field deflects the electrically charged particles in the wind. It uses wire loops, and runs a static current through them instead of applying a static voltage.

All these designs maneuver, though the mechanisms are different.

Magnetic sails bend the path of the charged protons that are in the solar wind. By changing the sails' attitudes, and the size of the magnetic fields, they can change the amount and direction of the thrust.

Sail Making

Materials

The most common material in current designs is a thin layer of aluminum coating on a polymer (plastic) sheet, such as aluminized 2 μm Kapton film. The polymer provides mechanical support as well as flexibility, while the thin metal layer provides the reflectivity. Such material resists the heat of a pass close to the Sun and still remains reasonably strong. The aluminum reflecting film is on the Sun side. The sails of Cosmos 1 were made of aluminized PET film (Mylar).

Eric Drexler developed a concept for a sail in which the polymer was removed. He proposed very high thrust-to-mass solar sails, and made prototypes of the sail material. His sail would use panels of thin aluminium film (30 to 100 nanometres thick) supported by a tensile structure. The sail would rotate and would have to be continually under thrust. He made and handled samples of the film in the laboratory, but the material was too delicate to survive folding, launch, and deployment. The design planned to rely on space-based production of the film panels, joining them to a deploy-able tension structure. Sails in this class would offer high area per unit mass and hence accelerations up to "fifty times higher" than designs based on deploy-able plastic films. The material developed for the Drexler solar sail was a thin aluminium film with a baseline thickness of 0.1 μm, to be fabricated by vapor deposition in a space-based system. Drexler used a similar process to prepare films on the ground. As anticipated, these films demonstrated adequate strength and robustness for handling in the laboratory and for use in space, but not for folding, launch, and deployment.

Research by Geoffrey Landis in 1998–1999, funded by the NASA Institute for Advanced Concepts, showed that various materials such as alumina for laser lightsails and carbon fiber for microwave pushed lightsails were superior sail materials to the previously standard aluminium or Kapton films.

In 2000, Energy Science Laboratories developed a new carbon fiber material that might be useful for solar sails. The material is over 200 times thicker than conventional solar sail designs, but it is so porous that it has the same mass. The rigidity and durability of this material could make solar sails that are significantly sturdier than plastic films. The material could self-deploy and should withstand higher temperatures.

There has been some theoretical speculation about using molecular manufacturing techniques to create advanced, strong, hyper-light sail material, based on nanotube mesh weaves, where the weave "spaces" are less than half the wavelength of light impinging on the sail. While such materials have so far only been produced in laboratory conditions, and the means for manufacturing such material on an industrial scale are not yet available, such materials could mass less than 0.1 g/m², making them lighter than any current sail material by a factor of at least 30. For comparison, 5 micrometre thick Mylar sail material mass 7 g/m², aluminized Kapton films have a mass as much as 12 g/m², and Energy Science Laboratories' new carbon fiber material masses 3 g/m².

The least dense metal is lithium, about 5 times less dense than aluminium. Fresh, unoxidized surfaces are reflective. At a thickness of 20 nm, lithium has an area density of 0.011 g/m². A high-performance sail could be made of lithium alone at 20 nm (no emission layer). It would have to be fabricated in space and not used to approach the Sun. In the limit, a sail craft might be constructed with a total areal density of around 0.02 g/m², giving it a lightness number of 67 and a_c of about 400 mm/s². Magnesium and beryllium are also potential materials for high-performance sails. These 3 metals can be alloyed with each other and with aluminium.

Reflection and Emissivity Layers

Aluminium is the common choice for the reflection layer. It typically has a thickness of at least 20 nm, with a reflectivity of 0.88 to 0.90. Chromium is a good choice for the emission layer on the face away from the Sun. It can readily provide emissivity values of 0.63 to 0.73 for thicknesses from 5 to 20 nm on plastic film. Usable emissivity values are empirical because thin-film effects dominate; bulk emissivity values do not hold up in these cases because material thickness is much thinner than the emitted wavelengths.

Fabrication

Sails are fabricated on Earth on long tables where ribbons are unrolled and joined to create the sails. Sail material needed to have as little weight as possible because it would require the use of the shuttle to carry the craft into orbit. Thus, these sails are packed, launched, and unfurled in space.

In the future, fabrication could take place in orbit inside large frames that support the sail. This would result in lower mass sails and elimination of the risk of deployment failure.

Operations

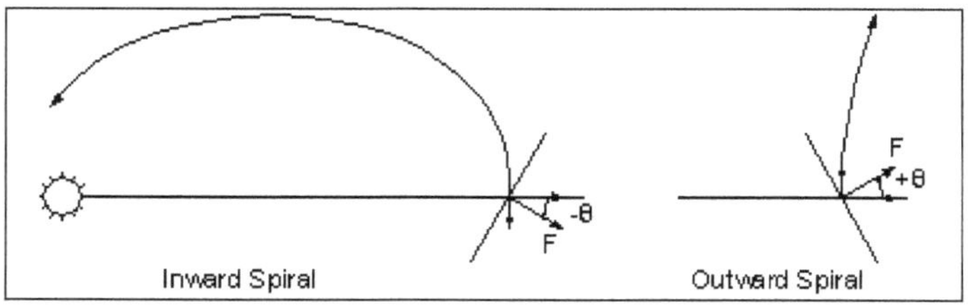

A solar sail can spiral inward or outward by setting the sail angle.

Changing Orbits

Sailing operations are simplest in interplanetary orbits, where altitude changes are done at low rates. For outward bound trajectories, the sail force vector is oriented forward of the Sun line, which increases orbital energy and angular momentum, resulting in the craft moving farther from the Sun. For inward trajectories, the sail force vector is oriented behind the Sun line, which decreases orbital energy and angular momentum, resulting in the craft moving in toward the Sun. It is worth noting that only the Sun's gravity pulls the craft toward the Sun—there is no analog to a sailboat's tacking to windward. To change orbital inclination, the force vector is turned out of the plane of the velocity vector.

In orbits around planets or other bodies, the sail is oriented so that its force vector has a component along the velocity vector, either in the direction of motion for an outward spiral, or against the direction of motion for an inward spiral.

Trajectory optimizations can often require intervals of reduced or zero thrust. This can be achieved by rolling the craft around the Sun line with the sail set at an appropriate angle to reduce or remove the thrust.

Swing-by Maneuvers

A close solar passage can be used to increase a craft's energy. The increased radiation pressure combines with the efficacy of being deep in the Sun's gravity well to substantially increase the energy for runs to the outer Solar System. The optimal approach to the Sun is done by increasing the orbital eccentricity while keeping the energy level as high as practical. The minimum approach distance is a function of sail angle, thermal properties of the sail and other structure, load effects on structure, and sail optical characteristics (reflectivity and emissivity). A close passage can result in substantial optical degradation. Required turn rates can increase substantially for a close passage. A sail craft arriving at a star can use a close passage to reduce energy, which also applies to a sail craft on a return trip from the outer Solar System.

A lunar swing-by can have important benefits for trajectories leaving from or arriving at Earth. This can reduce trip times, especially in cases where the sail is heavily loaded. A swing-by can also be used to obtain favorable departure or arrival directions relative to Earth.

A planetary swing-by could also be employed similar to what is done with coasting spacecraft, but good alignments might not exist due to the requirements for overall optimization of the trajectory.

The following table lists some example concepts using beamed laser propulsion as proposed by the physicist Robert L. Forward:

Mission	Laser Power	Vehicle Mass	Acceleration	Sail Diameter	Maximum Velocity (% of the speed of light)
1. Flyby – Alpha Centauri, 40 years					
outbound stage	65 GW	1 t	0.036 g	3.6 km	11% @ 0.17 ly
2. Rendezvous – Alpha Centauri, 41 years					
outbound stage	7,200 GW	785 t	0.005 g	100 km	21% @ 4.29 ly

deceleration stage	26,000 GW	71 t	0.2 g	30 km	21% @ 4.29 ly
3. Manned – Epsilon Eridani, 51 years (including 5 years exploring star system)					
outbound stage	75,000,000 GW	78,500 t	0.3 g	1000 km	50% @ 0.4 ly
deceleration stage	21,500,000 GW	7,850 t	0.3 g	320 km	50% @ 10.4 ly
return stage	710,000 GW	785 t	0.3 g	100 km	50% @ 10.4 ly
deceleration stage	60,000 GW	785 t	0.3 g	100 km	50% @ 0.4 ly

Interstellar travel catalog to use photogravitational assists for a full stop.

Name	Travel time (yr)	Distance (ly)	Luminosity ($L_\odot$)
Sirius A	68.90	8.58	24.20
α Centauri A	101.25	4.36	1.52
α Centauri B	147.58	4.36	0.50
Procyon A	154.06	11.44	6.94
Vega	167.39	25.02	50.05
Altair	176.67	16.69	10.70
Fomalhaut A	221.33	25.13	16.67
Denebola	325.56	35.78	14.66
Castor A	341.35	50.98	49.85
Epsilon Eridiani	363.35	10.50	0.50

- Successive assists at α Cen A and B could allow travel times to 75 yr to both stars.
- Lightsail has a nominal mass-to-surface ratio (σ_{nom}) of 8.6×10^{-4} gram m^{-2} for a nominal graphene-class sail.
- Area of the Lightsail, about 10^5 m^2 = $(316\ m)^2$.
- Velocity up to 37,300 km s^{-1} (12.5% c).

ELECTRIC SOLAR WIND SAIL

An electric sail (also known as an electric solar wind sail or an E-sail) is a proposed form of spacecraft propulsion using the dynamic pressure of the solar wind as a source of thrust. It creates a "virtual" sail by using small wires to form an electric field that deflects solar wind protons and extracts their momentum. The idea was first conceptualised by Pekka Janhunen in 2006 at the Finnish Meteorological Institute.

Academy of Finland has been funding electric sail development since 2007.

To test the technology, a new European Union-backed electric sail study project was announced by the FMI in December 2010. The EU funding contribution was 1.7 million euros. Its goal was to build laboratory prototypes of the key components, it involved five European countries and ended in November 2013. In the EU evaluation, the project got the highest marks in its category. An attempt was made to test the working principles of the electric sail in low Earth orbit in the Estonian nanosatellite ESTCube-1 (2013-2015), but there was a technical failure and the attempt was unsuccessful. The piezoelectric motor used to unfurl the sail failed to turn the reel. In subsequent ground-based testing, a likely reason for the failure was found in a slipring contact which was likely physically damaged by launch vibration.

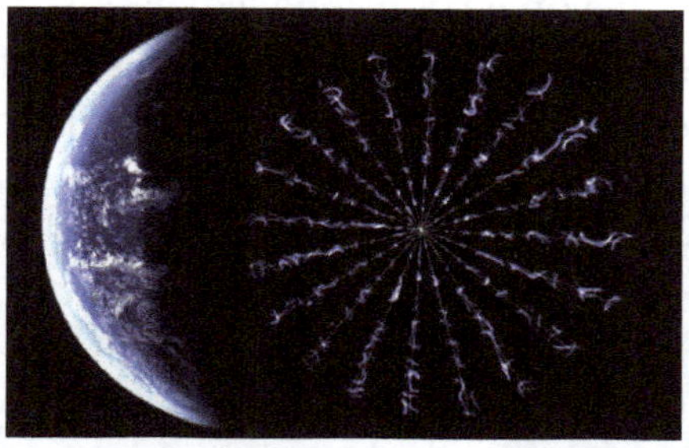

The Heliopause Electrostatic Rapid Transit System (HERTS) is a spacecraft concept using an electric sail.

An international research team that includes Janhunen received funding through a 2015 NIAC Phase II solicitation for further development at NASA's Marshall Space Flight Center. Their research project is called 'Heliopause Electrostatic Rapid Transit System' (HERTS). The Heliopause Electrostatic Rapid Transit System (HERTS) concept is currently being tested. For HERTS, it might take only 10 to 15 years to make the trip of over 100 astronomical units (15 billion kilometers). In the HERTS concept, multiple, 20 kilometer or so long, 1 millimeter thin, positively charged wires would be extended from a rotating spacecraft.

A new satellite launched in June 2017, the Finnish Aalto-1 nanosatellite, currently in orbit, will test the electric sail for deorbiting in 2019.

In 2017, Academy of Finland granted Centre of Excellence funding for 2018-2025 to a team that includes Janhunen and members from universities, to establish a Finnish Centre of Excellence in Research of Sustainable Space.

Principles of Operation and Design

The electric sail consists of a number of thin, long and conducting tethers which are kept in a high positive potential by an onboard electron gun. The positively charged tethers deflect solar wind protons, thus extracting momentum from them. Simultaneously they attract electrons from the solar wind plasma, producing an electron current. The electron gun compensates for the arriving electric current.

One way to deploy the tethers is to rotate the spacecraft, using centrifugal force to keep them stretched. By fine-tuning the potentials of individual tethers and thus the solar wind force individually, the spacecraft's attitude can be controlled.

E-sail missions can be launched at almost any time with only minor variations in travel time. By contrast, conventional slingshot missions must wait for the planets to reach a particular alignment.

The electric solar wind sail has little in common with the traditional solar sail. The E-sail gets its momentum from the solar wind ions, whilst a photonic sail is propelled by photons. Thus, the available pressure is only about 1% of photon pressure; however, this may be compensated by the simplicity of scale-up. In the E-sail, the part of the sail is played by straightened conducting tethers (made of wires) which are placed radially around the host ship. The wires are electrically charged and thus an electric field is created around the wires. The electric field of the wires extends a few dozen metres into the surrounding solar wind plasma. The penetration distance depends on the solar wind plasma density and it scales as the plasma Debye length. Because the solar wind electrons affect the electric field (similarly to the photons on a traditional solar sail), the effective electric radius of the tethers is based on the electric field that is generated around the tether rather than the actual tether itself. This fact also makes it possible to manoeuvre by regulating the tethers' electric charge.

Artist's rendering of ESTCube-1, launched in May 2013, which was intended to be the first satellite to test an electric sail.

A full-sized sail would have 50–100 straightened tethers with a length of about 20 km each. Compared to a reflective solar light sail, another propellantless deep space propulsion system, the electric solar wind sail could continue to accelerate at greater distances from the Sun, still developing thrust as it cruises toward the outer planets. By the time it reaches the ice giants, it may have accumulated as much as 20 km/s velocity, which is on par with the New Horizons probe, but without gravity assists.

In order to minimise damage to the thin tethers from micrometeoroids, the tethers would be formed from multiple strands, 25–50 micrometers in diameter, welded together at regular intervals. Thus, even if one wire were severed, a conducting path along the full length of the braided wire would remain in place. The feasibility of using ultrasonic welding was demonstrated at the University of Helsinki in January 2013.

Intrinsic limitations

Almost all Earth-orbiting satellites are inside Earth's magnetosphere. However, the electric sail cannot be used inside planetary magnetospheres because the solar wind does not penetrate them, allowing only slower plasma flows and magnetic fields. Instead, inside a planetary magnetosphere, the electric sail may function as a brake, allowing deorbiting of satellites.

Like for other solar sail technologies, while modest variation of the thrust direction can be achieved by inclining the sail, the thrust vector always points more or less radially outward from the Sun. It has been estimated that maximum operational inclination would be 60°, resulting in a thrusting angle of 30° from the outward radial direction. However, like with the sails of a ship, tacking could be used for changing the trajectory. Interstellar ships approaching a sun might use solar wind flow for braking.

Applications

- Fast missions (>50 km/s or 10 AU/year) out of the Solar system and heliosphere with small or modest payload.
- As a brake for a small interstellar probe which has been accelerated to high speed by some other means such as laser lightsail.
- Inward-spiralling missions to study the Sun at a closer distance.
- Two-way missions to inner Solar System objects such as asteroids.
- Off-Lagrange point solar wind monitoring spacecraft for predicting space weather with a longer warning time than 1 hour.

Fast Missions to Planet Uranus

Janhunen et al. have proposed a mission to Uranus powered by an electric sail. The mission could reach its destination in about the same time that the earlier Galileo space probe required to arrive at Jupiter, just over one fourth as far away. Galileo took 6 years to reach Jupiter at a cost of $1.6 billion, while Cassini-Huygens took 7 years to get to Saturn and cost almost as much. The sail is expected to consume 540 watts, producing about 0.5 newtons accelerating the craft by about 1 mm/s^2. The craft would reach a velocity of about 20 km/s by the time it reaches Uranus, 6 years after launch. The downside is that the electric sail cannot be used as a brake, so the craft arrives at a speed of 20 km/s, limiting the missions to flybys or atmospheric entry missions. Braking would require a conventional chemical rocket.

The proposed craft has three parts: the E-sail module with solar panels and reels to hold the wires; the main body, including chemical thrusters for adjusting trajectory en route and at destination and communications equipment; and a research module to enter Uranus's atmosphere and make measurements for relay to Earth via the main body.

References

- Perlin, John (2005). "Late 1950s – Saved by the Space Race". SOLAR EVOLUTION – The History of Solar Energy. The Rahus Institute. Retrieved 2007-02-25
- About-powersat: powersat.com, Retrieved 25 June, 2019
- Lee, Dillon (2008). "A Celebration of the Legacy of Physics at Dartmouth". Dartmouth Undergraduate Journal of Science. Dartmouth College. Retrieved 2009-06-11
- Solar-powersat-satellite-concept: permanent.com, Retrieved 26 July, 2019
- Kopp, G.; Lean, J. L. (2011). "A new, lower value of total solar irradiance: Evidence and climate significance". Geophysical Research Letters. 38 (1): n/a. Bibcode:2011georl..38.1706K. Doi:10.1029/2010GL045777
- Solar-powersat-satellite-space: permanent.com, Retrieved 27 August, 2019
- "Advanced Electric Propulsion System successfully tested at NASA's Glenn Research Center - spaceflight Insider". Www.spaceflightinsider.com. Retrieved 2018-07-28

Environmental Impacts of Solar Power

CHAPTER 7

Solar power has both positive and negative impact upon the environment. The major negative impacts of solar power are habitat loss due to land and water use, and the use of hazardous materials. This chapter closely examines these environmental impacts of solar power as well as the effect of photovoltaic systems and solar farms upon the environment.

ENVIRONMENTAL IMPACTS OF PHOTOVOLTAIC SYSTEMS

Undoubtedly, the sun is an incredible energy resource, allowing the generation of electricity seemingly without toxic pollutants or an effect on global warming. Some argue that photovoltaics are the cleanest form of electricity generation and few go as far as to claim that it is the only form with no effect on the environment at all. This notion however is flawed, as solar power does have significant and multidimensional environmental impacts, during the construction, operation and decommissioning phase.

The major environmental impacts of solar power are associated with:

1. The use of land.
2. The use of water.
3. The use of natural resources.
4. The use of hazardous materials.
5. The life-cycle global warming emissions.
6. The visual impact.

The scale and technology of the photovoltaic system has a direct effect on the level of each aforementioned impact. For example, building-integrated photovoltaics (BIPV) may require zero or nearly zero land use but may adversely affect the aesthetics of the site.

Land Use

Larger utility-scale photovoltaic electricity generation plants raise concerns about land degradation, loss of cultivable land, even that of habitat loss. Depending on the technology, site topography and location, estimates indicate that the generation of utility scale PV systems requires 12.000 m^2 to 40.000 m^2 per MW. Wind turbines require similarly large areas but there is greater opportunity to share land with other facilities, such as for agricultural use.

Studies suggest that the impact that utility scale PV systems have on land use may be minimized by placing them at location where the land quality is very low. These sites may be brownfields, abandoned mines or existing transportation/transmission corridors. Large-scale PV systems are generally suitable for use in isolated, deserted or other low land quality areas.

Utility-scale PV installation situated on barren land of very low value.

Medium-scale installations can also be integrated on large commercial buildings or as sound insulation barriers near sites of interest, such as hospitals, or between highways and residential/commercial areas.

Medium-scale PV installation acting as a highway sound barrier. Brenner Motorway, outside the town of Marano, Italy.

Small scale and building integrated PV arrays are usually installed on existing man-made structures and thus have minimal impact on the use of land. It is interesting to note that the net area of land made unavailable for other uses by PV power plants may be significantly less than for coal or nuclear energy, when all associated land uses, including mining, processing, etc. are taken into account.

Water Use

The use of water becomes a problematic issue with concentrating solar thermal plants that, like any thermal energy generation plant, requires water for cooling. As the sites that have the highest potential for such plants usually are those with dry climates, careful evaluation of the trade-offs is essential. Photovoltaic panels however seemingly do not use water for the generation of electricity. However, water is required for their maintenance, in order for the systems to maintain their maximum performance. Specifically, water is necessary for the cleaning of the panels, the amount of which greatly varies depending on the location of the system. Furthermore, as with most manufacturing processes, water is also required during the manufacture of photovoltaic panels and other components that a photovoltaic system requires.

Cleaning of a small-scale PV installation by jetting water.

Although it is not direct use of water, large utility-scale PV installations may have a severe impact on the hydrological resources at the installation site. They may reduce the rate of groundwater recharge, the filtration of the air and rainwater from pollutants and even increase the likelihood of flooding. This is especially likely if the installation forced the removal of trees on a sloping hillside. Still, if they are to be compared to any conventional form of energy, the effect of PV systems on these impacts is greatly beneficial.

Natural Resources Use

Aside from the production of current photovoltaic panels being an energy intensive process, it also requires very large quantities of bulk materials. Very large quantities of common minerals are required for the production of PV panels, such as iron, copper and aluminum. Aluminum and copper ores are not used by coal, petroleum or gas-fired energy generation stations at all, or are used in insignificant quantities. Iron is used in relatively large quantities by all conventional energy stations but still PV systems are estimated to require far greater quantities per produced kWh compared to all conventional forms of energy, including even coalfired installations.

Even though these materials are recyclable, the immense mineral depletion numbers should not be ignored. Studies report that 3.3 gr and 1.2 gr of iron and aluminum (Bauxite) ore are necessary per produced kWh. Such ures command the establishment of effective recycling programs if PV systems are to become a major energy contributor.

Also, even though the required quantities are small, scarce materials such as Telluride, Indium, Cadmium and Gallium. Some, such as Cadmium, are toxic as well. Acquiring these materials can be a costly and difficult process, especially when those are toxic.

Hazardous Materials

The manufacturing process of photovoltaic panels and associated components (e.g. inverters) contains a number of hazardous materials. The release of these hazardous materials to the environment is frequently considered to be the most critical negative environmental impact of both large and small PV systems. Most are used to clean and purify the semiconductor surface of photovoltaic cells. These chemicals are similar to those used in the general semiconductor industry and usually include:

- Hydrochloric acid.
- Sulfuric acid.
- Nitric acid.
- Hydrogen fluoride.
- 1,1,1-trichloroethane.
- Acetone.

The amount and type of chemicals used depends on the type of cell, the amount of cleaning that is needed, and the size of silicon wafer. More advanced technologies also tend to broaden the list of hazardous materials present during their manufacture. For example, thin-film photovoltaics contain more toxic materials than traditional silicon-based cells, including gallium arsenide, copper indium gallium diselenide (CIGS), and cadmium telluride. In normal operation, cadmium telluride cells do not pose any environmental risk. Nevertheless, in a fire, and only in a fire, poisonous gases can be produced. Copper indium diselenide cells are considered less potentially hazardous than cadmium telluride cells. If not handled and disposed of properly, these materials could pose serious environmental or public health threats. However, manufacturers have a strong financial incentive to ensure that these highly valuable and often rare materials are recycled rather than thrown away.

Still, if compared to conventional forms of energy, the life-cycle of PV systems involves far less hazardous materials. For example, Cadmium emissions are nearly 300 times greater in the case of coal-fired energy stations. The emissions of common pollutants which are dangerous for the well-being of humans, such as NO_x and SO_2, are orders of magnitude lower than any conventional form of energy, with the exception of nuclear energy. The combustion emissions of conventional fuel-fired power stations also include pollutants that are associated with cancer, reproductive, and cardiovascular effects. In that respect, PV systems are greatly superior to any combustion-driven power station.

The manufacturing process of photovoltaic cells is also a known health hazard for workers, who face the risk of inhaling silicon dust.

Life-cycle Global Warming Emissions

Undoubtedly, there are virtually zero global warming emissions during the operation of photovoltaic systems. However, emissions are generated through almost every other stage of the photovoltaic life cycle. Some of these stages include the manufacturing of the photovoltaic panels and other associated parts, their transportation, the installation and site alteration procedures, maintenance processes, even the recycling and decommissioning of defunct installations. Primary energy consumption for the manufacturing of photovoltaic panels has been one of the most serious concerns of researchers for decades. An earlier review paper summarized many of these studies and the author specifically noted the significant disagreement between researchers. As the author reports, there were studies indicating that the energy consumption required for the production of one m^2 of multicrystalline silicon (mc-Si) modules may be as low as 667 kWh or as high as 2115 kWh. Similarly, the reported energy consumption for monocrystalline silicon (mn-Si) panels was between 1470 kWh and 4580 kWh per m^2. Thin-film technologies are much less energy-intensive, generally requiring about a quarter of the energy required to make an equivalent mc-Si panel. The review however covered older studies, before 2000. As such, they were based on assumptions involving older manufacturing methods and silicon processing. In addition, it is should be noted that all similar studies have to assume the estimated production of the PV system, which obviously differs from site to site. As such, studies that were performed for areas with low solar irradiation obviously report less optimistic results.

Most estimates of life-cycle emissions for photovoltaic systems are between 30 and 80 grams of carbon dioxide equivalent per generated kWh, depending on the energy generation mix of the area/country. There are studies indicating that newer photovoltaic technologies, such as Cadmium Telluride (Cd-Te) panels, can go as low as 24 grams/kWh, rivalling the nuclear energy baseline. These emissions are almost entirely associated with the production and transportation of the materials required for a photovoltaic installation, i.e. the panels and the power electronics. The emissions associated with the materials of conventional power plants are a very low percentage of their total emissions per generated kWh, as the bulk of their greenhouse gas (GHG) emissions come from their operation. Still, even if their operation emissions are not accounted at all, solely the production and the transportation of their fuel greatly surpasses the GHG emissions of photovoltaic panels per kWh.

Visual Impact

Another important environmental impact is the visual impact that PV systems may have on buildings and areas. The visual intrusion of a building-integrated PV system can be very high but can also be addressed with relative ease. Careful architectural designing and materials can introduce the photovoltaic panels as architectural elements, complementing the appearance of the building instead of degrading it. It is not necessary for the architect to hide panels and components entirely, only to implement them as part of the overall building design. From the side of PV panel manufacturers, modern designs can easily be developed specifically for their installation in populated areas and on buildings, with colored versions already in the market for the few years. Colored PV panels are frequently used to simulate the appearance of ceramic roof tiles or typical glass. Retrofitting PV systems on roofs and large surfaces of older buildings usually is a viable solution as well, with the exception of buildings that cannot be visually altered, i.e. those which are of historical or cultural importance. Furthermore, thin film panels could easily replace the mirrors and glass surfaces of large buildings. As thin film panels have a lower solar transmittance than glass, they could also be used as a means to reduce the cooling load of buildings, offering shading or even heat extraction. However, the architect always has to consider the effects of glare as well. The replacement of glass surfaces may very well be the only method applicable to historical buildings and buildings with high cultural value, where visual modifications of the structure itself are impossible.

Pfizer-University of Granada-Junta de Andalucía Center for Genomic and Oncologic Research (GENyO) ventilated photovoltaic façade designed by Onyx Solar. It generates about 32 MWh/annum.

Utility-scale installations are a more complicated matter. The installation of a utility-scale installation at or near areas of natural beauty, tourist attractions, archeological sites, ecological areas and other similar locations is problematic and should be avoided. Large installations need to be properly sited, usually outside residential and commercial areas as well. Abandoned areas and low-quality land usually are the most suitable sites for such installations.

Polycrystalline PV panels retrofitted to cover the facade of a commercial building.
They were used to replace mirrors.

POSITIVE AND NEGATIVE ENVIRONMENTAL IMPACTS OF SOLAR PANELS

Are solar panels really that green? Much has been discussed on this topic from both sides of the political spectrum.

Many argue that solar panels are not that green since they require energy to manufacture and sometimes use harmful chemical during the processing. Here we will examine the positive and negative effects of solar energy and what can be done moving forward.

Let's start by saying that solar power isn't perfect. Like everything in life, there are upsides and downsides especially when trying to generate enough energy to sustain a reasonable quality of life. Yes, vast amounts of energy are required and yes, many solar modules use toxic chemicals during the manufacturing process. These two irrefutable facts do not equate to solar panels having a negative net impact though.

Lifetime Carbon Emissions from Solar

To understand the positive and negative environmental impacts of solar power, it is important to realize what goes into producing a functional solar panel.

First the raw material must be mined, this is predominantly Quartz which is processed into silicon. Aluminum, and copper or silver are also key materials involved which must be mined, or obtained from recycled sources, but mostly they are mined due to the increased expansion of the PV industry in the last 10 years.

Following the mining of raw materials, the quartz must be processed into electronic-grade silicon. This process involves heating the quartz in a high temperature furnace and reacting it with various chemicals.

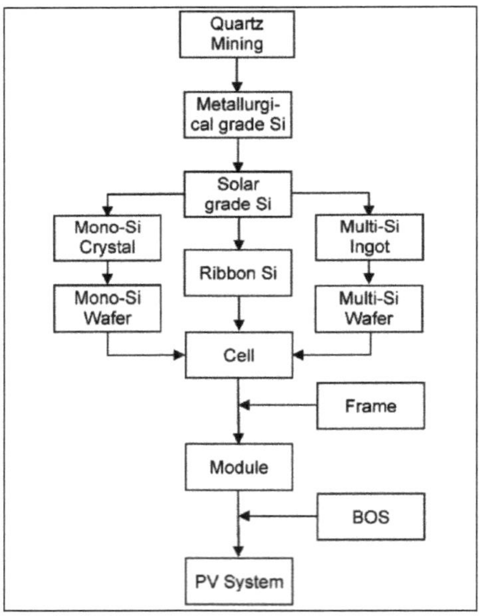

Solar Panel Manufacturing Process.

The lifetime carbon emissions from solar are heavily front loaded. After the installation process, solar modules produce emission-free energy for 30+ years. The question to focus on now is:

> "Does the clean energy generated from solar panels offset the negative impacts during the mining and manufacturing process?"

Carbon Footprint of Solar Panel Manufacturing

In a whitepaper from the Brookhaven National Laboratory PV Environmental Research Centre, the authors outline the environmental impacts of solar panel manufacturing.

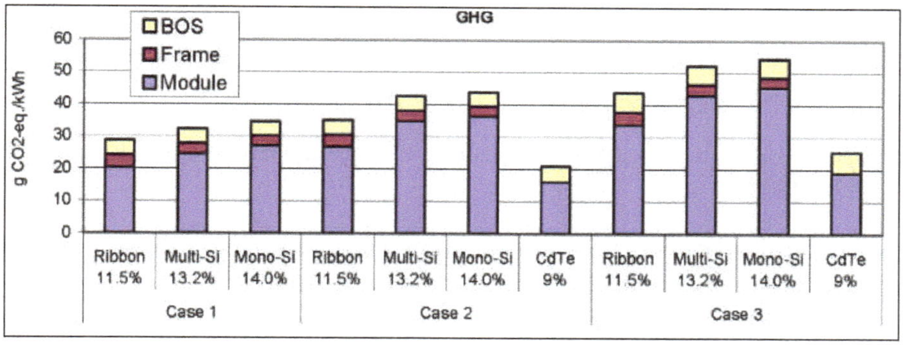

In the figure, Case 1 is a representation of CrystalClear and Ecoinvent Database set, Case 2 is a Union of Coordination of Transmission of Electricity and Ecoinvent database set, and Case 3 is US grid mixture and Franklin database set. For simplification we will observe the average of the 3 data sets for polycrystalline silicon (Multi-Si).

When the three multi-silicon values (~32, ~42, ~52 gCo_2/kWh) are average, the result is an average of 42 gCo_2-eq/kWh of energy generated. This value is approximately in line with more recent data from the World Nuclear Association which estimates the lifetime emissions output of solar PV to be ~70tCo_2-eq/GWh (g/kWh is an equal ratio to a t/GWh).

Negative Environmental Impacts Solar Panels

Solar power is not without its downsides. Let's examine them here:

1. Energy - The manufacturing of solar panels requires more energy up front to produce than other forms of energy generation. This is due to the manufacturing process required to transform raw minerals into usable photovoltaics. Quartz must be processed, and cleaned and then manufactured with other components which may come from different facilities (aluminum, copper etc.) to produce a single solar module whereas coal is mined, cleaned and burned on a mass scale typically in a single location. Heating the quartz during the processing stage requires very high heat. Manufacturing requires combining multiple materials with incredible precision to produce high efficiency panels. All of this requires lots of up-front energy.

2. Chemicals – To produce solar-grade silicon, semi-conductors processing typically involves hazardous chemicals. Depending on the solar panel manufacturer, these chemicals may or may not be disposed properly. The Silicon Valley Toxics Coalition (SVTC) presents an annual report on the environmental records of major solar manufacturers. Like every industry, there are companies leading by example, and others which cut corners to save money. Not every company will dump chemicals, or recycle their byproducts improperly, but there are a few bad apples out there fueling the anti-green bandwagon.

3. Recycling – Although solar panel recycling has not become a major issue yet, it will in the coming decades as solar panels need to be replaced. Technology does however tend to keep up with demand. The automobile recycling industry did not begin until the auto industry was well underway. The same can be said for other recycling industries such as bottle depots, biofuel, and electronics. Companies such as PV Cycle are already thinking ahead to mitigate this problem.

These are the major environmental concerns surrounding the PV industry but overall, the downsides are minimal compared to the downsides associated with common forms of generating energy such as coal and natural gas.

Although solar panels do require a great amount of energy up front, they do have a very short energy payback time (EPBT).

Toxic chemicals are required during the manufacturing stage, but hazardous materials are required for almost every type of energy generation. Coal must be cleaned with chemicals and burned. Natural gas must be extracted and burned. Nuclear energy itself requires a highly radioactive material. Wind turbines use metal which must be mined, and processed too. No energy form is perfect, but some are better than others. To tell the difference we will explore the carbon emission intensity of each of these fuel types below.

Energy Payback Time of Solar Panels

The energy payback period for solar power depends on your location and what the typical form of energy generation is. Only when compared to the form of energy displaced, can the payback period be observed. If your home is powered 100% by wind and hydro energy, then replacing that with solar will not provide much environmental benefit (although there will be other benefits such as relieving pressure on the electric grid, and providing you ownership of your energy). But if your home is powered by coal, then solar power will have a very fast environmental payback period.

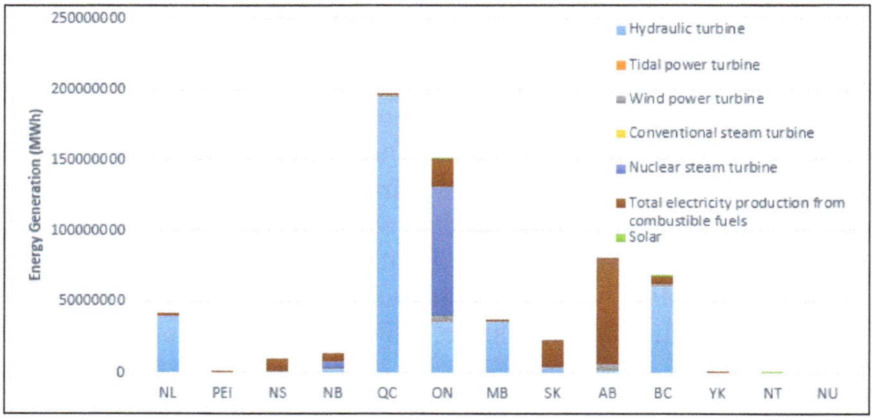

Canada's energy generation by province and fuel type.

Provinces such as Nova Scotia, Saskatchewan and Alberta would benefit from solar power since most energy in these provinces come from fossil fuels.

To provide some general context, the U.S. Department of Energy's National Renewable Energy Laboratory (NREL) provides some noteworthy data. Poly-Crystalline solar panels have an energy payback period of 2 to 3.5 years. This value includes the manufacturing of the module, frame and the balance of the system components.

Carbon Emission Intensity

The values observed here are paramount to understanding the lifetime emission profile from solar energy and other energy generating methods.

Although all data has a fair degree of variance, which is expected for a rapidly progressing industry, the general trends in comparison remain identical.

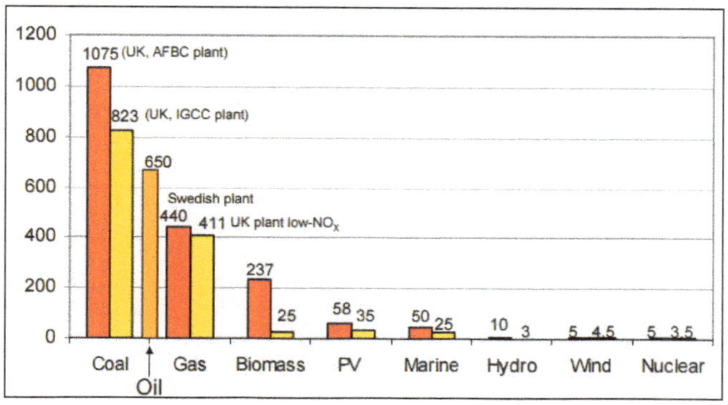

Both the emission intensity profiles show solar PV being 3x to 10x more environmentally friendly over a 30 year time span compared to energy generated from coal.

Solar PV does require heavy amounts of energy up front to mine and manufacture the materials, but when that emission is dispersed over a 30 year generation profile the emissions/kWh are much more favourable. This key metric must be considered when evaluating the environmental impacts of solar panels.

Solar power is not perfect, but overall it provides a positive net environmental impact and excellent long-term financials.

The energy required to create a solar panel will be recouped after just 2 to 4 years. Even considering the manufacturing and processing stage of solar, the emissions generated are 3x to 10x less than generating the same amount of energy from fossil fuels.

Naturally, the benefits will vary depending on the energy generation and solar irradiance of you location but overall solar panels provide a positive net impact.

ENVIRONMENTAL EFFECTS OF SOLAR FARM

Solar energy is touted to be one of the greenest, cleanest sources of electricity. But you might be surprised to know that it is a potential threat to the environment. Well, no doubt, solar energy offers significant benefits compared to conventional forms of electricity production, but solar farms tend to have a deep environmental impact, from land degradation to habitat destruction and ecological imbalance. Let's look at some of the issues that solar farms present to decide whether "green energy is clean energy."

The Risk of Emissions

Well, we all know that solar energy is one of the cleanest sources of energy, with no risk of global warming emissions. However, you might be surprised to learn that there are emissions associated with different stages of the solar lifecycle. To begin with, it is important to understand the solar panel production process. It all begins with the mining and processing of the raw material. Quartz, silver, copper, and aluminum are involved in the mining process. Quartz then undergoes further processing with hazardous chemicals in high-temperature furnaces to produce electronic grade silicon.

The major point of concern in the production process of silicon is the leftover silver content, which is considered a potential environmental hazard. The large-scale production of solar modules could amount to the depletion of silver resources.

The Risk of Solar Waste Disposal

Not much has been done on this front. The disposal of the chemicals involved in the production depends on the manufacturer, raising serious questions about the environmental safety. While there are responsible solar manufacturers that choose to safely dispose of solar waste, there are others that aren't too concerned about environmental safety and don't hesitate to cut corners to save money. Such irresponsible acts by solar manufacturers can cause more environmental harm than good.

Negative Effects Solar Energy Consumption

The production of solar modules consumes a high amount of energy. Additionally, the resultant water and hazardous byproducts are released into the environment during the manufacturing process.

The initial stages of the solar panel manufacturing process consume more energy compared to any other form of electricity production. The manufacturing process involves the transformation of raw mineral into usable photovoltaics. The different components involved in the processing, mining, production and cleaning of raw material may come from different facilities unlike conventional sources of energy production, where everything takes place at a single spot.

Besides, during the manufacturing process, the heating of quartz requires a lot of upfront energy. This can quickly amount to energy loss.

The Dark Side: Water Wastage

Like other conventional energy generation plants, some solar plants may require large amounts of water for cooling turbines and cleaning solar collectors. Increase in water demand could strain the available water resources. This may amount to the loss of large volumes of groundwater in arid locations, which may further impact the local wildlife that depends on these water resources. Besides, the use of chemicals, such as dust suppressants and dielectric fluids, may cause surface contamination or degradation of groundwater.

Environmental Impact: Habitat Degradation

The installation of solar panel farms requires clearing vast tracts of land, which could affect existing land uses, including agriculture, grazing, and mineral production. Clearing and grading of land may amount to habitat destruction and pose a risk of soil erosion and compaction. This may further interfere with the existing use of land and alteration of drainage channels.

Unfortunately, habitat destruction for the wildlife could result in a large-scale displacement of wild species. There have been reports of numerous bird deaths in California's Ivanpah Solar Generating System. Reports blame the deaths on the solar farm's mirrors, claiming that heat reflected from them burned off the bird's wings.

The impact on one wildlife species can destroy the entire ecosystem since one species is dependent on another. When one is removed, the entire ecosystem of that species, comprised of plants, animals, birds, and marine life, can starve to death. The reason is that the habitat becomes less livable for the species that are dependent on each other for survival.

The conservation of wildlife habitat and diversity is a major cause for concern among environmentalists.

Recycling is a Headache

One area that needs greater attention is recycling of solar waste. Recycling is likely to become a major issue in the coming times, as solar waste can prove to be a big burden on the environment. Recycling technology for reusing silicon from solar cells is not in place yet, though research has been underway to prove that the manufacturing of a solar panel from recycled components consumes significantly less amount of energy than solar module production from scratch.

Solar panel manufacturers need to delve deeper into the issue to mitigate the problem and focus on recycling.

The Hidden Aspect: Climate Change

While the world sees a climate change fix in solar energy, there is a debate among environmentalists whether this form of energy is worth it. There are concerns about the rampant development of solar panel farms and its impact on the broader environment. A report published claims that there could be severe temperature reductions beneath solar installations by as much as 5 degrees Celsius in the hot season. This could affect plant growth in the surrounding areas.

However, a deeper understanding of the environmental impact of solar installations could enlighten farmers on microclimate changes and how to better utilize the vacant land under panels. Farmers may need to think deeper about choosing crops that can survive in low temperatures created by the shade of the solar panels.

NEGATIVE ENVIRONMENTAL IMPACTS OF SOLAR ENERGY

Extensive fossil fuel consumption in almost all human activities has led to some undesirable phenomena such as atmospheric and environmental pollutions, which have not been experienced before in known human history. With the objective of emerging as a global leader in solar energy by increasing the solar generation capacity to 20 GW by 2022, the National Solar Mission sets ambitious target of harnessing around 5,000 trillion kWh of solar energy in a phased manner through the use of appropriate technology and by developing substantial solar manufacturing capacity within the country. Solar energy technologies are essential components of a sustainable energy future. Nevertheless, solar energy systems have some certain negative impacts on the environment just like any other energy system.

Land Use and Thermal Pollution

Solar cells (photovoltaics) have miscellaneous impacts on the natural ecosystem. These impacts are related to some specific factors, like the area and the topography of land that bwould be covered, sensitive ecosystems, and biodiversity. The application of the solar cells in cultivable land can cause possible harm on the land's productive areas. Large-scale utilization of the land also effects thermal balance of the area by absorbing more energy by the earth than otherwise would be reflected by the surface back to space. A serious solar power application needs to utilize square kilometers of desert area. Thermal balance of this land space can certainly be effected by such an application. Also, additional heat might destroy a few species living in this kind of harsh environment.

Discharge of Pollutants

Solar cells do not emit any pollutants during their operations. But solar cell modules contain some toxic substances, and there is a potential risk of releasing these chemicals to the environment during a fire. Necessary precautions should be taken for emergency situations like fire. The sibility of an accidental release of the chemicals of the solar cell modules to soil and groundwater poses a great threat for the environment.

Visual Impacts

There will be some visual impacts depending on the type of the scheme and the surroundings of the solar cells. Especially for applications on the buildings, solar cells can be used as a cladding material that could be integrated into the building during the construction phase. Solar cell applications after the construction phase of the buildings might cause negative visual impacts. Solar cell utilization should be planned at the architectural phase and fitted to the building to minimize visual pollution. For the other application areas, proper sitting and design are important factors, especially for large solar cell applications. Another important factor about the control of the visual impacts is the use of color. Enough care should be taken for the usage of proper colors while assembling the solar cell modules.

An example of solar panlels not causing visual impact.

Impacts on Natural Resources

Solar cells have some negative impacts on the environment during their production phase like many other systems. The energy needed for the production of solar energy systems is still produced in conventional methods today. Some toxic chemical substances used during the production phase are produced as a by-product. Especially, the solar cell batteries pose a threat on natural resources by having a short lifespan and containing heavy metals such as cadmium.

Air Pollution

Solar cells do not emit any substances to the air during operation. But there could be some emissions during manufacturing and transport. The emissions associated with the transport of the modules are insignificant when compared with the emissions associated with the manufacture.

Noise Intrusion

Solar cells do not make a noise during operation. But during the construction phase, there will be a little noise as usual in other construction activities.

PERMISSIONS

All chapters in this book are published with permission under the Creative Commons Attribution Share Alike License or equivalent. Every chapter published in this book has been scrutinized by our experts. Their significance has been extensively debated. The topics covered herein carry significant information for a comprehensive understanding. They may even be implemented as practical applications or may be referred to as a beginning point for further studies.

We would like to thank the editorial team for lending their expertise to make the book truly unique. They have played a crucial role in the development of this book. Without their invaluable contributions this book wouldn't have been possible. They have made vital efforts to compile up to date information on the varied aspects of this subject to make this book a valuable addition to the collection of many professionals and students.

This book was conceptualized with the vision of imparting up-to-date and integrated information in this field. To ensure the same, a matchless editorial board was set up. Every individual on the board went through rigorous rounds of assessment to prove their worth. After which they invested a large part of their time researching and compiling the most relevant data for our readers.

The editorial board has been involved in producing this book since its inception. They have spent rigorous hours researching and exploring the diverse topics which have resulted in the successful publishing of this book. They have passed on their knowledge of decades through this book. To expedite this challenging task, the publisher supported the team at every step. A small team of assistant editors was also appointed to further simplify the editing procedure and attain best results for the readers.

Apart from the editorial board, the designing team has also invested a significant amount of their time in understanding the subject and creating the most relevant covers. They scrutinized every image to scout for the most suitable representation of the subject and create an appropriate cover for the book.

The publishing team has been an ardent support to the editorial, designing and production team. Their endless efforts to recruit the best for this project, has resulted in the accomplishment of this book. They are a veteran in the field of academics and their pool of knowledge is as vast as their experience in printing. Their expertise and guidance has proved useful at every step. Their uncompromising quality standards have made this book an exceptional effort. Their encouragement from time to time has been an inspiration for everyone.

The publisher and the editorial board hope that this book will prove to be a valuable piece of knowledge for students, practitioners and scholars across the globe.

INDEX

CPSIA information can be obtained
at www.ICGtesting.com
Printed in the USA
LVHW060854030122
707669LV00002B/152

9 781641 165945